Survival Analysis Using SAS®

A Practical Guide Second Edition

Paul D. Allison

§.sas.

CONTENTS

Chapter 7 Analysis of Tied or Discrete Data with PROC LOGISTIC   235

Chapter 8 Heterogeneity, Repeated Events, and Other Topics   257

Chapter 9 A Guide for the Perplexed   289

Appendix 1 Macro Programs   293

Appendix 2 Data Sets   299

VI

PREFACE

When the first edition of Survival Analysis Using SAS was published in 1995, my goal was to provide an accessible, data-based introduction to methods of survival analysis, one that focused on methods available in SAS and that also used SAS for the examples. The success of that book confirmed my belief that statistical methods are most effectively taught by showing researchers how to implement them with familiar software using real data.

Of course, the downside of a software-based statistics text is that the software often changes more rapidly than the statistical methodology. In the 15 years that the first edition of the book has been in print, there have been so many changes to the features and syntax of SAS procedures for survival analysis that a new edition has been long overdue. Indeed, I have been working on this second edition for several years, but got partially sidetracked by a four-year term as department chair. So, it's a great relief that I no longer have to warn potential readers about out-of-date SAS code.

Although the basic structure and content of the book remain the same, there are numerous small changes and several large changes. One global change is that all the figures use ODS Graphics. Here are the other major changes and additions:

- Chapter 3, "Estimating and Comparing Survival Curves with PROC LIFETEST." This chapter documents some major enhancements to the STRATA statement, which now offers several alternative tests for comparing survivor functions. It also allows for pairwise comparisons and for adjustment of p-values for multiple comparisons. In the first edition, I demonstrated the use of a macro called SMOOTH, which I had written to produce smoothed graphs of hazard function. That macro is no longer necessary, however, because the PLOTS option (combined with ODS Graphics) can now produce smoothed hazard functions using a variety of methods.

- Chapter 4, "Estimating Parametric Regression Models with PROC LIFEREG." This chapter now includes a section on the PROBPLOT command, which offers graphical methods to evaluate the fit of each model. The last section introduces the new BAYES statement, which (as the name suggests) makes it possible to do a Bayesian analysis of any of the parametric models using MCMC methods.

- Chapter 5, "Estimating Cox Regression Models with PROC LIFEREG." The big change here is the use of the counting process syntax as an alternative method for handling time-dependent covariates. When I wrote the first edition, the counting process syntax had just been introduced, and I did not fully appreciate its usefulness for handling predictor variables that vary over time. Another new topic is the use of the ASSESS statement to evaluate the proportional hazards assumption. Finally, there is a section that describes the BAYES statement for estimating Cox models and piecewise exponential models.
- Chapter 6, "Competing Risks." This chapter now contains a section on cumulative incidence functions, which is a popular alternative approach to competing risks.
- Chapter 7, "Analysis of Tied or Discrete Data with the LOGISTIC Procedure." The first edition also used the PROBIT and GENMOD procedures to do discrete time analysis. But, PROC LOGISTIC has been enhanced to the point where the other procedures are no longer needed for this application.
- Chapter 8, "Heterogeneity, Repeated Events, and Other Topics." For repeated events and other kinds of clustered data, the WLW macro that I described in the first edition has been superseded by the built-in option COVSANDWICH. In this chapter, I also describe the use of the new GLIMMIX procedure to estimate random-effects models for discrete time data.

Please note that I use the following convention for presenting SAS programs. All words that are part of the SAS language are shown in uppercase. All user-specified variable names and data set names are in lowercase. In the main text itself, both SAS keywords and user-specified variables are in uppercase.

I am most grateful to my editor, George McDaniel, for his patient persistence in getting me to finish this new edition.

CHAPTER **1**
Introduction

WHAT IS SURVIVAL ANALYSIS?

Survival analysis is a class of statistical methods for studying the occurrence and timing of events. These methods are most often applied to the study of deaths. In fact, they were originally designed for that purpose, which explains the name. That name is somewhat unfortunate, however, because it encourages a highly restricted view of the potential applications of these methods. Survival analysis is extremely useful for studying many different kinds of events in both the social and natural sciences, including disease onset, equipment failures, earthquakes, automobile accidents, stock market crashes, revolutions, job terminations, births, marriages, divorces, promotions, retirements, and arrests. Because these methods have been adapted—and sometimes independently discovered—by researchers in several different fields, they also go by several different names: event history analysis (sociology), reliability analysis (engineering), failure time analysis (engineering), duration analysis (economics), and transition analysis (economics). These different names don't imply any real difference in techniques, although different disciplines may emphasize slightly different approaches. Because survival analysis is the name that is most widely used and recognized, it is the name I use here.

This book is about doing survival analysis with SAS. I have also written an introduction to survival analysis that is not oriented toward a specific statistical package (Allison, 1984), but I prefer the approach taken here. To learn any kind of statistical analysis, you need to see how it's actually performed in some detail. And to do that, you must use a particular computer program. But which one? Although I have performed survival analysis with many different statistical packages, I am convinced that SAS currently has the most comprehensive set of full-featured procedures for doing survival analysis. When I compare SAS with any of its competitors in this area, I invariably find some crucial capability that SAS has but that the other package does not. When you factor in the extremely powerful tools that SAS provides for data management and manipulation, the choice is clear. On the other hand, no statistical package can do everything, and some methods of survival analysis are not available in SAS. I occasionally mention such methods, but the predominant emphasis in this book is on those things that SAS can actually do.

I don't intend to explain every feature of the SAS procedures discussed in this book. Instead, I focus on those features that are most widely used, most potentially useful, or most likely to cause problems and confusion. It's always a good idea to check the official SAS documentation or online help file.

WHAT IS SURVIVAL DATA?

Survival analysis was designed for longitudinal data on the occurrence of events. But what is an event? Biostatisticians haven't written much about this question because they have been overwhelmingly concerned with deaths. When you consider other kinds of events, however, it's important to clarify what is an event and what is not. I define an *event* as a qualitative change that can be situated in time. By a *qualitative change*, I mean a transition from one discrete state to another. A marriage, for example, is a transition from the state of being unmarried to the state of being married. A promotion consists of the transition from a job at one level to a job at a higher level. An arrest can be thought of as a transition from, say, two previous arrests to three previous arrests.

To apply survival analysis, you need to know more than just who is married and who is not married. You need to know *when* the change occurred. That is, you should be able to situate the event in time. Ideally, the transitions occur virtually instantaneously, and you know the exact times at which they occur. Some transitions may take a little time, however, and the exact time of onset may be unknown or ambiguous. If the event of interest is a political revolution, for example, you may know only the year in which it began. That's all right so long as the interval in which the event occurs is short relative to the overall duration of the observation.

You can even treat changes in *quantitative* variables as events if the change is large and sudden compared to the usual variation over time. A fever, for example, is a sudden, sustained elevation in body temperature. A stock market crash could be defined as any single-day loss of more than 20 percent in some market index. Some researchers also define events as occurring when a quantitative variable crosses a threshold. For example, a person is said to have fallen into poverty when income goes below some designated level. This practice may not be unreasonable when the threshold is an intrinsic feature of the phenomenon itself or when the threshold is legally mandated. But I have reservations about the application of survival methods when the threshold is arbitrarily set by the researcher. Ideally, statistical models should reflect the process generating the observations. It's hard to see how such arbitrary thresholds can accurately represent the phenomenon under investigation.

For survival analysis, the best observation plan is prospective. You begin observing a set of individuals at some well-defined point in time, and you follow them for some substantial period of time, recording the times at which the events of interest occur. It's not necessary that every individual experience the event. For some applications, you may also want to distinguish between different kinds of events. If the events are deaths, for example, you might record the cause of death. Unlike deaths, events like arrests, accidents, or promotions are repeatable; that is, they may occur two or more times to the same individual. While it is definitely desirable to observe and record multiple occurrences of the same event, you need specialized methods of survival analysis to handle these data appropriately.

You can perform survival analysis when the data consist *only* of the times of events, but a common aim of survival analysis is to estimate causal or predictive models in which the risk of an event depends on covariates. If this is the goal, the data set must obviously contain measurements of the covariates. Some of these covariates, like race and sex, may be constant over time. Others, like income, marital status, or blood pressure, may vary with time. For time-varying covariates, the data set should include as much detail as possible on their temporal variation.

Survival analysis is frequently used with *retrospective* data in which people are asked to recall the dates of events like marriages, child births, and promotions. There is nothing intrinsically wrong with this approach as long as you recognize the potential limitations. For one thing, people may make substantial errors in recalling the times of events, and they may forget some events entirely. They may also have difficulty providing accurate information on time-dependent covariates. A more subtle problem is that the sample of people who are actually interviewed may be a biased subsample of those who may have been at risk of the event. For example, people who have died or moved away will not be included. Nevertheless, although prospective data are certainly preferable, much can be learned from retrospective data.

4   ## WHY USE SURVIVAL ANALYSIS?

Survival data have two common features that are difficult to handle with conventional statistical methods: *censoring* and *time-dependent covariates* (sometimes called *time-varying explanatory variables*). Consider the following example, which illustrates both these problems. A sample of 432 inmates released from Maryland state prisons was followed for one year after release (Rossi et al., 1980). The event of interest was the first arrest. The aim was to determine how the occurrence and timing of arrests depended on several covariates (predictor variables). Some of these covariates (like race, age at release, and number of previous convictions) remained constant over the one-year interval. Others (like marital and employment status) could change at any time during the follow-up period.

How do you analyze such data using conventional methods? One possibility is to perform a logistic regression analysis with a dichotomous dependent variable: arrested or not arrested. But this analysis ignores information on the timing of arrests. It's natural to suspect that people who are arrested one week after release have, on average, a higher propensity to be arrested than those who are not arrested until the 52nd week. At the least, ignoring that information should reduce the precision of the estimates.

One solution to this problem is to make the dependent variable the length of time between release and first arrest and then estimate a conventional linear regression model. But what do you do with the persons who were not arrested during the one-year follow-up? Such cases are referred to as *censored*. There are a couple of obvious ad-hoc methods for dealing with censored cases, but neither method works well. One method is to discard the censored cases. That method might work reasonably well if the proportion of censored cases is small. In our recidivism example, however, fully 75 percent of the cases were not arrested during the first year after release. That's a lot of data to discard, and it has been shown that large biases may result. Alternatively, you could set the time of arrest at one year for all those who were not arrested. That's clearly an underestimate, however, and some of those ex-convicts may *never* be arrested. Again, large biases may occur.

Whichever method you use, it's not at all clear how a time-dependent variable like employment status can be appropriately incorporated into either the logistic model for the occurrence of arrests or the linear model for the timing of arrests. The data set contains information on whether each person was working full time during each of the 52 weeks of follow-up. You could, I suppose, estimate a model with 52 indicator (dummy) variables for employment status. Aside from the computational awkwardness and statistical inefficiency of such a procedure, there is a more fundamental problem: all the employment indicators for weeks *after* an arrest might be *consequences* of the arrest rather than causes. In particular, someone who is jailed after an arrest is not likely to be working full time in subsequent weeks. In short, conventional methods don't offer much hope for dealing with either censoring or time-dependent covariates.

By contrast, all methods of survival analysis allow for censoring, and many also allow for time-dependent covariates. In the case of censoring, the trick is to devise a procedure that combines the information in the censored and uncensored cases in a way that produces consistent estimates of the parameters of interest. You can easily accomplish this by the method of maximum likelihood or its close cousin, partial likelihood. Time-dependent covariates can also be incorporated with these likelihood-based methods. Later chapters explain how you can usefully apply these methods to the recidivism data.

APPROACHES TO SURVIVAL ANALYSIS

One of the confusing things about survival analysis is that there are so many different methods: life tables, Kaplan-Meier estimators, exponential regression, log-normal regression, proportional hazards regression, competing risks models, and discrete-time methods, to name only a few. Sometimes these methods are complementary. Life tables have a very different purpose than regression models, for example, and discrete-time methods are designed for a different kind of data than continuous-time methods. On the other hand, it frequently happens that two or more methods may seem attractive for a given application, and the researcher may be hard-pressed to find a good reason for choosing one over another. How do you choose between a log-normal regression model (estimated with the LIFEREG procedure) and a proportional hazards model (estimated with the PHREG procedure)? Even in the case of discrete-time versus continuous-time methods, there is often considerable uncertainty about whether time is best treated as continuous or discrete. One of the aims of this book is to help you make intelligent decisions about which method is most suitable for your particular application. SAS/STAT software contains six procedures that can be used for survival analysis. Here's an overview of what they do:

LIFETEST   is primarily designed for univariate analysis of the timing of events. It produces life tables and graphs of survival curves (also called *survivor functions*). Using several methods, this procedure tests whether survival curves are the same in two or more groups. PROC LIFETEST also tests for associations between event times and time-constant covariates, but it does not produce estimates of parameters.

LIFEREG   estimates regression models with censored, continuous-time data under several alternative distributional assumptions. PROC LIFEREG allows for several varieties of censoring, but it does not allow for time-dependent covariates.

PHREG   uses Cox's partial likelihood method to estimate regression models with censored data. The model is less restrictive than the models in PROC LIFEREG, and the estimation method allows for time-dependent covariates. PROC PHREG handles both continuous-time and discrete-time data.

LOGISTIC   is designed for general problems in categorical data analysis, but it is effective and flexible in estimating survival models for discrete-time data with time-dependent covariates.

GENMOD   estimates the same discrete-time survival models as PROC LOGISTIC, but it can handle repeated events using generalized estimating equation (GEE) methods.

NLMIXED   estimates random-effects (mixed) parametric models for repeated events.

All of these procedures can be used to estimate competing risks models that allow for multiple kinds of events, as described in Chapter 6, "Competing Risks."

WHAT YOU NEED TO KNOW

I have written this book for the person who wants to analyze survival data using SAS but who knows little or nothing about survival analysis. The book should also be useful if you are already knowledgeable about survival analysis and simply want to know how to do it with SAS. I assume that you have a good deal of practical experience with ordinary least squares regression analysis and that you are reasonably familiar with the assumptions of the linear model. There is little point in trying to estimate and interpret regression models for survival data if you don't understand ordinary linear regression analysis.

You do not need to know matrix algebra, although I sometimes use the vector notation $\boldsymbol{\beta}\mathbf{x} = \beta_1 x_1 + \beta_2 x_2 + ... \beta_k x_k$ to simplify the presentation of regression models. A basic knowledge of limits, derivatives, and definite integrals is helpful in following the discussion of hazard functions, but you can get by without those tools. Familiarity with standard properties of logarithms and exponentials is essential, however. Chapters 4 and 5 each contain a more technical section that you can skip without loss of continuity. (I note this at the beginning of the section.) Naturally, the more experience you have with SAS/STAT and the SAS DATA step, the easier it will be to follow the discussion of SAS statements. On the other hand, the syntax for most of the models considered here is rather simple and intuitive, so don't be intimidated if you are a SAS neophyte.

COMPUTING NOTES

Most of the examples in this book were executed on a Dell laptop running Windows Vista at 1.80 GHz with 1 GB of memory, using SAS 9.2. Occasionally, I report comparisons of computing times for different procedures, options, sample sizes, and program code. These comparisons should only be taken as rough illustrations, however, because computing time depends heavily on both the software and hardware configuration.

8

CHAPTER **2**
Basic Concepts of Survival Analysis

INTRODUCTION

In this chapter, I discuss several topics that are common to many different methods of survival analysis:

- censoring, a nearly universal feature of survival data
- common ways of representing the probability distribution of event times, especially the survivor function and the hazard function
- choice of origin in the measurement of time, a tricky but important issue that is rarely discussed in the literature on survival analysis
- the basic data structure required for most computer programs that perform survival analysis.

CENSORING

Not all survival data contain censored observations, and censoring may occur in applications other than survival analysis. Nevertheless, because censored survival data are so common and because censoring requires special treatment, it is this topic more than anything else that unifies the many approaches to survival analysis.

Censoring comes in many forms and occurs for many different reasons. The most basic distinction is between *left censoring* and *right censoring*. An observation on a variable T is right censored if all you know about T is that it is greater than some value c. In survival analysis, T is typically the time of occurrence for some event, and cases are right censored because observation is terminated before the event occurs. Thus, if T is a person's age at death (in years), you may know only that $T > 50$, in which case the person's death time is right censored at age 50. This notion of censoring is not restricted to event times. If you know only that a person's income is greater than \$75,000 per year, then income is right censored at \$75,000.

Symmetrically, left censoring occurs when all you know about an observation on a variable T is that it is *less* than some value. Again, you can apply this notion to any sort of variable, not just an event time. In the context of survival data, left censoring is most likely to occur when you begin observing a sample at a time when some of the individuals may have already experienced the event. If you are studying menarche (the onset of menstruation), for example, and you begin following girls at age 12, you may find that some of them have already begun menstruating. Unless you can obtain information on the starting date for those girls, the age of menarche is said to be left censored at age 12. (In the social sciences, *left censoring* often means something quite different. Observations are said to be left censored if the *origin time*, not the event time, is known only to be less than some value. According to the definitions used here, such observations are actually right censored.)

In both the natural and social sciences, right censoring is far more common than left censoring, and most computer programs for survival analysis do not allow for left censored data. Note, however, that the LIFEREG procedure *will* handle left censoring as well as *interval censoring*. Interval censoring combines both right and left censoring. An observation on a variable T is interval censored if all you know about T is that $a < T < b$, for some values of a and b. For survival data, this sort of censoring is likely to occur when observations are made at infrequent intervals, and there is no way to get retrospective information on the exact timing of events. Suppose, for example, that a sample of people is tested annually for HIV infection. If a person who was not infected at the end of year 2 is then found to be infected at the end of year 3, the time of infection is interval censored between 2 and 3. When all observations are interval censored and the intervals are equally spaced, it is often convenient to treat such data as discrete-time data, which is the subject of Chapter 7, "Analysis of Tied or Discrete Data with PROC LOGISTIC."

I won't say anything more about left censoring and interval censoring until Chapter 4, "Estimating Parametric Regression Models with PROC LIFEREG." The rest of this section is about the various patterns of right-censored data and the possible mechanisms generating such data. The distinctions are important because some kinds of censoring are unproblematic, while other kinds require some potentially dubious assumptions.

The simplest and most common situation is depicted in Figure 2.1. For concreteness, suppose that this figure depicts some of the data from a study in which all persons receive heart surgery at time 0 and are followed for 3 years thereafter. The horizontal axis represents time. Each of the horizontal lines labeled A through E represents a different person. An X indicates that a death occurred at that point in time. The vertical line at 3 is the point at which we stop following the patients. Any deaths occurring at time 3 or earlier are observed and, hence, those death times are uncensored. Any deaths occurring after time 3 are not observed, and those death times are censored at time 3. Therefore, persons A, C, and D have uncensored death times, while persons B and E have right-censored death times. Observations that are censored in this way are referred to as *singly Type I censored*.

Figure 2.1   *Singly Right-Censored Data*

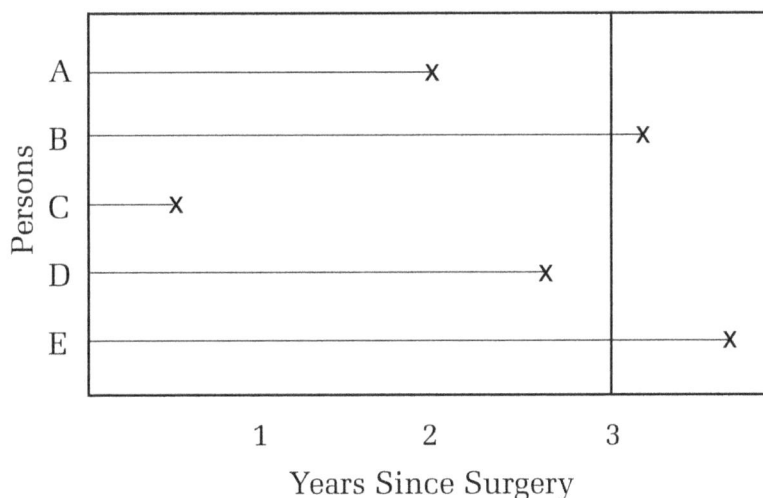

Type I means that the censoring time is fixed (that is, under the control of the investigator), and *singly* refers to the fact that all the observations have the same censoring time. Even observations that are not censored are said to have a censoring time, in this case 3 years. It's just that their death times did not exceed their censoring time. Of course, censoring times can also vary across individuals. For example, you might want to combine data from two experiments, one with observation terminating after 3 years and another with observation terminating after 5 years. This is still Type I censoring, provided the censoring time is fixed by the design of the experiment.

Type II censoring occurs when observation is terminated after a prespecified number of events have occurred. Thus, a researcher running an experiment with 100 laboratory rats may decide that the experiment will stop when 50 of them have died. This sort of censoring is uncommon in the social sciences.

Random censoring occurs when observations are terminated for reasons that are *not* under the control of the investigator. There are many possible reasons why this might happen. Suppose you are interested in divorces, so you follow a sample of couples for 10 years beginning with the marriage, and you record the timing of all divorces. Clearly, couples that are still married after 10 years are censored by a Type I mechanism. But for some couples, either the husband or the wife may die before the 10 years are up. Some couples may move out of state or to another country, and it may be impossible to contact them. Still other couples may refuse to participate in the study after, say, 5 years. These kinds of censoring are depicted in Figure 2.2, where the O for couples B and C indicates that observation is censored at that point in time. Regardless of the subject matter, nearly all prospective studies end up with some cases that didn't make it to the maximum observation time for one reason or another.

Figure 2.2 *Randomly Censored Data*

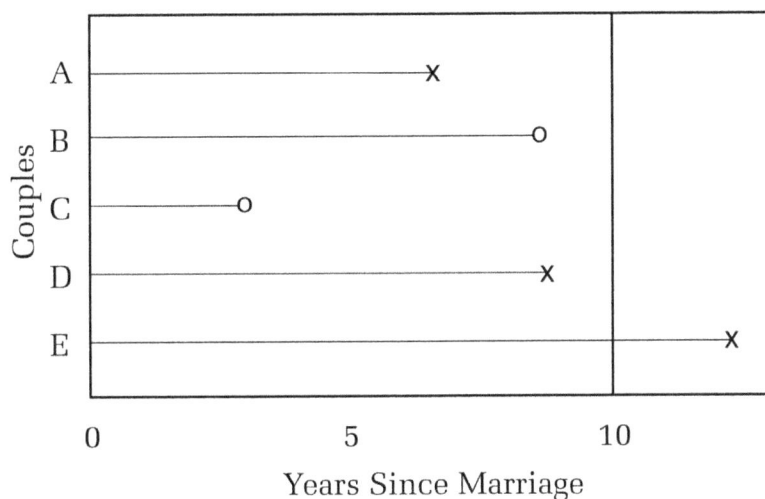

Random censoring can also be produced when there is a single termination time, but entry times vary *randomly* across individuals. Consider again the example in which people are followed from heart surgery until death. A more likely scenario is one in which people receive heart surgery at various points in time, but the study has to be terminated on a single date (say, December 31, 2010). All persons still alive on that date are considered censored, but their survival times from surgery will vary. This censoring is considered random because the entry times are typically not under the control of the investigator.

Standard methods of survival analysis do not distinguish among Type I, Type II, and random censoring. They are all treated as generic right-censored observations. Why make the distinctions, then? Well, if you have only Type I or Type II censoring, you're in good shape. The maximum likelihood and partial likelihood methods discussed in this book handle these types of censoring with no appreciable bias. Things are not so simple with random censoring, however. Standard methods require that random censoring be noninformative. Here's how Cox and Oakes (1984) describe this condition:

> A crucial condition is that, conditionally on the values of any explanatory variables, the prognosis for any individual who has survived to c_i should not be affected if the individual is censored at c_i. That is, an individual who is censored at c should be representative of all those subjects with the same values of the explanatory variables who survive to c (p. 5).

The best way to understand this condition is to think about possible violations. Suppose you follow a cohort of new graduate students to see what factors affect how long it takes them to get a Ph.D. Many students drop out before completing the degree, and these observations are randomly censored. Unfortunately, there is good reason to suspect that those who drop out are among those who would take a long time to finish if they stayed until completion. This is called *informative censoring*. In the divorce example mentioned earlier, it is plausible that those couples who refuse to continue participating in the study are more likely to be experiencing marital difficulties and, hence, at greater risk of divorce. Again, the censoring is informative (assuming that measured covariates do not fully account for the association between dropout and marital difficulty).

Informative censoring can, at least in principle, lead to severe biases, but it is difficult in most situations to gauge the magnitude or direction of those biases. In the case of graduate student drop out, where the censored cases are likely to be those who would have had long times to the event, one consequence of censoring is to underestimate the median survival time.

An easy solution to censoring that is random because of random entry times is to include entry time as a covariate in a regression model. This solution should work well in many situations, but it can lead to computational difficulties if a large proportion of the observations is censored.

Unfortunately, there is no statistical test for informative censoring versus noninformative censoring. The best you can do is a kind of sensitivity analysis that is described in Chapter 8, "Heterogeneity,

Repeated Events, and Other Topics." Aside from that, there are three important lessons here:

1. First, in designing and conducting studies, you should do everything possible to reduce the amount of random censoring. You can't rely on statistical methods to completely adjust for such censoring.

2. Second, you should make an effort to measure and include in the model any covariates that are likely to affect the rate of censoring.

3. Third, in studies with high levels of random censoring, you may want to place less confidence in your results than calculated confidence intervals indicate.

DESCRIBING SURVIVAL DISTRIBUTIONS

All of the standard approaches to survival analysis are *probabilistic* or *stochastic*. That is, the times at which events occur are assumed to be realizations of some random process. It follows that *T*, the event time for some particular individual, is a random variable having a probability distribution. There are many different models for survival data, and what often distinguishes one model from another is the probability distribution for *T*. Before looking at these different models, you need to understand three different ways of describing probability distributions.

Cumulative Distribution Function

One way that works for all random variables is the *cumulative distribution function*, or *c.d.f.* The c.d.f. of a variable *T*, denoted by $F(t)$, is a function that tells us the probability that the variable will be less than or equal to any value *t* that we choose. Thus, $F(t) = \Pr\{T \le t\}$. If we know the value of *F* for every value of *t*, then we know all there is to know about the univariate distribution of *T*. In survival analysis, it is more common to work with a closely related function called the *survivor function*, defined as $S(t) = \Pr\{T > t\} = 1 - F(t)$. If the event of interest is a death, the survivor function gives the probability of surviving beyond *t*. Because *S* is a probability, we know that it is bounded by 0 and 1. And because *T* cannot be negative, we know that $S(0) = 1$. Finally, as *t* gets larger, *S* never increases (and usually decreases). Within these restrictions, *S* can have a wide variety of shapes.

Chapter 3, "Estimating and Comparing Survival Curves with PROC LIFETEST," explains how to estimate survivor functions using life-table and Kaplan-Meier methods. Often, the objective is to compare survivor functions for different subgroups in a sample. If the survivor function for one group is always higher than the survivor function for another group, then the first group clearly lives longer than the second group. If survivor functions cross, however, the situation is more ambiguous.

Probability Density Function

When variables are continuous, another common way of describing their probability distributions is the *probability density function,* or *p.d.f.* This function is defined as

$$f(t) = \frac{dF(t)}{dt} = -\frac{dS(t)}{dt} \qquad (2.1)$$

That is, the p.d.f. is just the derivative or slope of the c.d.f. Although this definition is considerably less intuitive than that for the c.d.f., it is the p.d.f. that most directly corresponds to our intuitive notions of distributional shape. For example, the familiar bell-shaped curve that is associated with the normal distribution is given by its p.d.f., not its c.d.f.

Hazard Function

For continuous survival data, the *hazard function* is actually more popular than the p.d.f. as a way of describing distributions. The hazard function is defined as

$$h(t) = \lim_{\Delta t \to 0} \frac{\Pr(t \leq T < t + \Delta t \mid T \geq t)}{\Delta t} \qquad (2.2)$$

Instead of $h(t)$, some authors denote the hazard by $\lambda(t)$ or $r(t)$. Because the hazard function is so central to survival analysis, it is worth taking some time to explain this definition. The aim of the definition is to quantify the instantaneous risk that an event will occur at time t. Because time is continuous, the probability that an event will occur at exactly time t is necessarily 0. But we *can* talk about the probability that an event occurs in the small interval between t and $t + \Delta t$. We also want to make this probability *conditional* on the individual surviving to time t. Why? Because if individuals have already *died* (that is, experienced the event), they are clearly no longer at risk of the event. Thus, we want to consider only those individuals who have made it to the beginning of the interval $[t, t + \Delta t)$. These considerations point to the numerator in equation (2.2): $\Pr(t \leq T < t + \Delta t \mid T \geq t)$.

The numerator is still not quite what we want, however. First, the probability is a nondecreasing function of Δt—the longer the interval, the more likely it is that an event will occur in that interval. To adjust for this, we divide by Δt, as in equation (2.2). Second, we want the risk for event occurrence at *exactly* time t, not in some interval beginning with t. So we shrink the interval down by letting Δt get smaller and smaller, until it reaches a limiting value.

The definition of the hazard function in equation (2.2) is similar to an alternative definition of the probability density function:

$$f(t) = \lim_{\Delta t \to 0} \frac{\Pr(t \le T < t + \Delta t)}{\Delta t} \tag{2.3}$$

The only difference is that the probability in the numerator of equation (2.3) is an unconditional probability, whereas the probability in equation (2.2) is conditional on $T \ge t$. For this reason, the hazard function is sometimes described as a *conditional density*. When events are repeatable, the hazard function is often referred to as the *intensity function*.

The survivor function, the probability density function, and the hazard function are equivalent ways of describing a continuous probability distribution. Given any one of them, we can recover the other two. The relationship between the p.d.f. and the survivor function is given directly by the definition in equation (2.1). Another simple formula expresses the hazard in terms of the p.d.f. and the survivor function:

$$h(t) = \frac{f(t)}{S(t)}. \tag{2.4}$$

Together, equations (2.4) and (2.1) imply that

$$h(t) = -\frac{d}{dt} \log S(t). \tag{2.5}$$

Integrating both sides of equation (2.5) gives an expression for the survivor function in terms of the hazard function:

$$S(t) = \exp\left[-\int_0^t h(u)du \right]. \tag{2.6}$$

Together with equation (2.4), this formula leads to

$$f(t) = h(t)\exp\left[-\int_0^t h(u)du \right]. \tag{2.7}$$

These formulas are extremely useful in any mathematical treatment of models for survival analysis because it is often necessary to move from one representation to another.

INTERPRETATIONS OF THE HAZARD FUNCTION

Before proceeding further, three clarifications need to be made:

- Although it may be helpful to think of the hazard as the instantaneous probability of an event at time t, it's not really a probability because the hazard can be greater than 1.0. This can happen because of the division by Δt in equation (2.1). Although the hazard has no upper bound, it cannot be less than 0.

- Because the hazard is defined in terms of a probability (which is never directly observed), it is itself an unobserved quantity. We may estimate the hazard with data, but that's only an estimate.

- It's most useful to think of the hazard as a characteristic of individuals, not of populations or samples (unless everyone in the population is exactly the same). Each individual may have a hazard function that is completely different from anyone else's.

The hazard function is much more than just a convenient way of describing a probability distribution. In fact, the hazard at any point t corresponds directly to intuitive notions of the risk of event occurrence at time t. With regard to numerical magnitude, the hazard is a dimensional quantity that has the form *number of events per interval of time*, which is why the hazard is sometimes called a *rate*. To interpret the value of the hazard, then, you must know the units in which time is measured. Suppose, for example, that I somehow know that my hazard for contracting influenza at some particular point in time is .015, with time measured in months. This means that if my hazard stays at that value over a period of one month, I would expect to contract influenza .015 times. Remember, this is not a probability. If my hazard was 1.3 with time measured in years, then I would expect to contract influenza 1.3 times over the course of a year (assuming that my hazard stays constant during that year).

To make this more concrete, consider a simple but effective way of estimating the hazard. Suppose that we observe a sample of 10,000 people over a period of one month, and we find 75 cases of influenza. If every person is observed for the full month, the total exposure time is 10,000 months. Assuming that the hazard is constant over the month and across individuals, an optimal estimate of the hazard is 75/10000=.0075. If some people died or withdrew from the study during the one-month interval, we have to subtract their *unobserved* time from the denominator.

The assumption that the hazard is constant may bother some readers because one thing known about hazards is that they can vary continuously with time. That's why I introduced the hazard *function* in the first place. Yet, this sort of hypothetical interpretation is one that is familiar to everyone. If we examine the statement "This car is traveling at 30 miles per hour," we are actually saying that "*If* the car continued at this constant speed for a period of one hour, it would travel 30 miles." But cars never maintain exactly the same speed for a full hour.

The interpretation of the hazard as the expected number of events in a one-unit interval of time is sensible when events are repeatable. But what about a nonrepeatable event like death? Taking the reciprocal of the hazard, $1/h(t)$, gives the expected length of time until the event occurs, again assuming that $h(t)$ remains constant. If my hazard for death is .018 per year at this moment, then I can expect to live another $1/.018 = 55.5$ years. Of course, this calculation assumes that everything about me and my environment stays exactly the same. Actually, my hazard of death will certainly increase (at an increasing rate) as I age. The reciprocal of the hazard is useful for repeatable events as well. If I have a constant hazard of .015 per month of contracting influenza, the expected length of time between influenza episodes is 66.7 months.

In thinking about the hazard, I find it helpful to imagine that each of us carries around hazards for different kinds of events. I have a hazard for accidental death, a hazard for coronary infarction, a hazard for quitting my job, a hazard for being sued, and so on. Furthermore, each of these hazards changes as conditions change. Right now, as I sit in front of my computer, my hazard for serious injury (one requiring hospitalization) is very low but not zero. The ceiling could collapse or my chair could tip over, for example. It surely goes up substantially as I leave my office and walk down the stairs. And it goes up even more when I get in my car and drive onto the expressway. Then it goes down again when I get out of my car and walk into my home.

This example illustrates the fact that the true hazard function for a specific individual and a specific event varies greatly with the ambient conditions. In fact, it is often a step function with dramatic increases or decreases as an individual moves from one situation to another. When we estimate a hazard function for a group of individuals, these micro-level changes typically cancel out so that we end up capturing only the gross trends with age or calendar time. On the other hand, by including changing conditions as time-dependent covariates in a regression model (Chapter 5, "Estimating Cox Regression Models with PROC PHREG"), we *can* estimate their effects on the hazard.

SOME SIMPLE HAZARD MODELS

We have seen that the hazard function is a useful way of describing the probability distribution for the time of event occurrence. Every hazard function has a corresponding probability distribution. But hazard functions can be extremely complicated, and the associated probability distributions may be rather esoteric. This section examines some rather simple hazard functions and discusses their associated probability distributions. These hazard functions are the basis for some widely employed regression models that are introduced briefly here.

The simplest function says that the hazard is constant over time: that is, $h(t) = \lambda$ or, equivalently, $\log h(t) = \mu$. Substituting this hazard into equation (2.6) and carrying out the integration implies that the survival function is $S(t) = e^{-\lambda t}$. Then, from equation (2.1), we get the p.d.f., $f(t) = \lambda e^{-\lambda t}$. This is the p.d.f. for the well-known exponential distribution with parameter λ. Thus, a constant hazard implies an exponential distribution for the time until an event occurs (or the time between events).

The next step up in complexity is to let the natural logarithm of the hazard be a linear function of time:

$$\log h(t) = \mu + \alpha t. \tag{2.8}$$

Taking the logarithm is a convenient and popular way to ensure that $h(t)$ is nonnegative, regardless of the values of μ, α, and t. Of course, we can rewrite the equation as

$$h(t) = \lambda \gamma^t \tag{2.9}$$

where $\lambda = e^{\mu}$ and $\gamma = e^{\alpha}$. This hazard function implies that the time of event occurrence has a *Gompertz* distribution. Alternatively, we can assume that

$$\log h(t) = \mu + \alpha \log t \tag{2.10}$$

which can be rewritten as

$$h(t) = \lambda t^{\alpha} \qquad\qquad (2.11)$$

with $\lambda = e^{\mu}$. This equation implies that the time of event occurrence follows a *Weibull* distribution.

Figures 2.3 and 2.4 show some typical hazard functions for the Weibull and Gompertz distributions. Both distributions have the exponential distribution as a special case when α is 0. When α is not 0, the hazard is either always increasing or always decreasing with time for both distributions. One difference between them is that, for the Weibull model, when $t = 0$, the hazard is either 0 or infinite. With the Gompertz model, on the other hand, the initial value of the hazard is just λ, which can be any nonnegative number.

Figure 2.3 *Typical Hazard Functions for the Weibull Distribution*

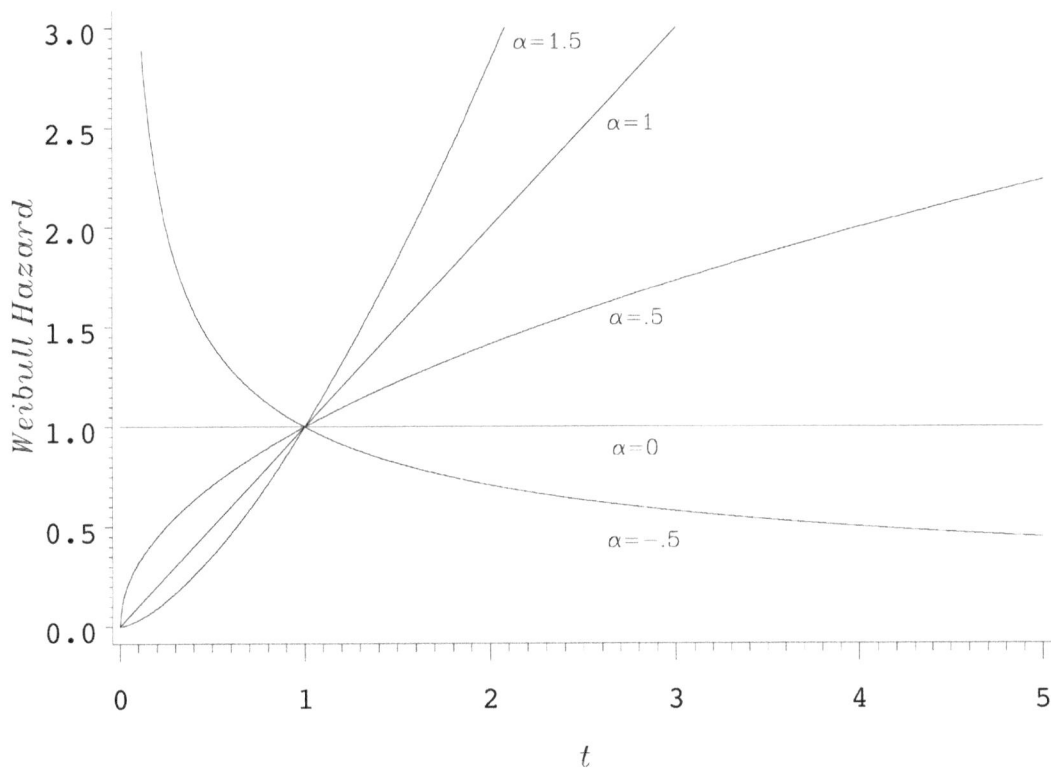

Figure 2.4   *Typical Hazard Functions for the Gompertz Distribution*

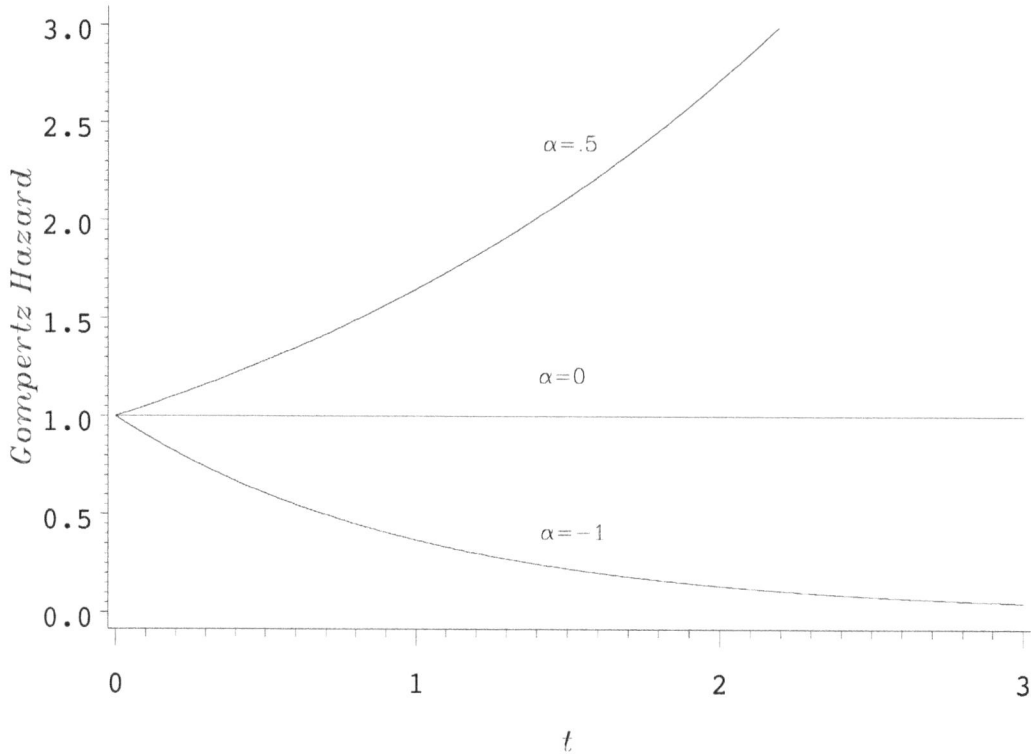

We can readily extend each of these models to allow for the influence of covariates (explanatory variables). Thus, if we have covariates $x_1, x_2,..., x_k$, we can write

$$\text{Exponential:}\quad \log h(t) = \mu + \beta_1 x_1 + \beta_2 x_2 + ... + \beta_k x_k$$

$$\text{Gompertz:}\quad \log h(t) = \mu + \alpha t + \beta_1 x_1 + \beta_2 x_2 + ... + \beta_k x_k \qquad (2.12)$$

$$\text{Weibull:}\quad \log h(t) = \mu + \alpha \log t + \beta_1 x_1 + \beta_2 x_2 + ... + \beta_k x_k$$

We can estimate the Weibull and exponential models (along with a number of other models) with PROC LIFEREG, as described in Chapter 4. The Gompertz model is not standardly available in SAS. All three models are members of a general class known as *proportional hazards models*. Chapter 5 explains how to use Cox's partial likelihood method to estimate the coefficients of the covariates for any proportional hazards model without having to specify exactly how the hazard depends on time.

The Origin of Time

All models for survival data are fundamentally concerned with the timing of events. In assigning a number to an event time, we implicitly choose both a scale and an origin. The scale is just the units in which time is measured: years, days, minutes, hours, or seconds. We have already seen that the numerical value of the hazard depends on the units of measurement for time. In practice, however, the choice of units makes little difference for the regression models discussed in later chapters. Because those models are linear in the *logarithm* of the hazard or event time, a change in the units of measurement affects only the intercept, leaving the coefficients unchanged.

The choice of origin (0 point) is more problematic, however, for three reasons:

1. First, it *does* make a difference—often substantial—in coefficient estimates and fit of the models.
2. Second, the preferred origin is sometimes unavailable, in which case you must use some proxy.
3. Third, many situations occur in which two or more possible time origins are available, but there is no unambiguous criterion for deciding among them.

Consider the problem of unavailability of the preferred origin. Many medical studies measure time of death as the length of time between the *point of diagnosis* and death. Most medical researchers prefer, if possible, to measure time from the point of infection or the onset of the disease. Because there is often wide variation in how long it takes before a disease is diagnosed, the use of diagnosis time as a proxy may introduce a substantial amount of random noise into the measurement of death times. A likely consequence is attenuation of coefficients toward 0. Worse yet, because variation in time of diagnosis may depend on such factors as age, sex, race, and social class, there is also the possibility of systematic bias. Thus, if African Americans tend to be diagnosed later than Caucasians, they will appear to have shorter times to death. Unfortunately, if the point of disease onset is unavailable (as it usually is), you cannot do much about this problem except to be aware of the potential biases that might result.

On the other hand, the point of disease onset may *not* be the ideal choice for the origin. If the risk of death depends heavily on treatment—

which cannot begin until the disease is diagnosed—then the point of diagnosis may actually be a better choice for the origin. This fact brings up the third issue. What criteria can be used to choose among two or more possible origins? Before attempting to answer that question, consider some of the possibilities:

- *Age.* Demographers typically study the age at death, implicitly using the individual's birth as the point of origin.
- *Calendar time.* Suppose we begin monitoring a population of deer on October 1, 2010, and follow them for one year, recording any deaths that occur in that interval. If we know nothing about the animals prior to the starting date, then we have no choice but to use the starting date as the origin for measuring death time.
- *Time since some other event.* In studying the determinants of divorce, it is typical to use the date of marriage as the origin time. Similarly, in studying criminal recidivism, the natural starting date is the date at which the convict is released from prison.
- *Time since the last occurrence of the same type of event.* When events are repeatable, it is common to measure the time of an event as the time since the most recent occurrence. Thus, if the event is a hospitalization, we may measure the length of time since the most recent hospitalization.

In principle, the hazard for the occurrence of a particular kind of event can be a function of *all* of these times or any subset of them. Nevertheless, the continuous-time methods considered in this book require a choice of a single time origin. (The discrete-time methods discussed in Chapter 7 are more flexible in that regard.) Although you can sometimes include time measurements based on other origins as covariates, that strategy usually restricts the choice of models and may require more demanding computation.

So how do you choose the principal time origin? Here are some criteria that researchers commonly use, although not always with full awareness of the rationale or the implications:

- *Choose a time origin that marks the onset of continuous exposure to risk of the event.* If the event of interest is a divorce, the natural time origin is the date of the marriage. Prior to marriage, the risk (or hazard) of divorce is 0. After marriage, the risk is some positive number. In the case of recidivism, a convict is not at risk of recidivating until he or she is actually released from prison, so the point of release is an obvious time origin.

This criterion is so intuitively appealing that most researchers instinctively apply it. But the justification is important because there are sometimes attractive alternatives. The most compelling argument for this criterion is that it automatically excludes earlier periods of time when the hazard is necessarily 0. If these periods are not excluded, and if they vary in length across individuals, then bias may result. (Chapter 5 shows how to exclude periods of zero hazard using PROC PHREG.)

Often this criterion is qualified to refer only to some subset of a larger class of events. For example, people are continuously at risk of death from the moment they are born. Yet, in studies of deaths due to radiation exposure, the usual origin is the time of first exposure. That is the point at which the individual is first exposed to risk of that particular kind of death—a death due to radiation exposure. Similarly, in a study of why some patients die sooner than others after cardiac surgery, the natural origin is the time of the surgery. The event of interest is then *death following cardiac surgery*. On the other hand, if the aim is to estimate the effect of surgery itself on the death rate among cardiac patients, the appropriate origin is time of diagnosis, with the occurrence of surgery as a time-dependent covariate.

- *In experimental studies, choose the time of randomization to treatment as the time origin.* In such studies, the main aim is usually to estimate the differential risk associated with different treatments. It is only at the point of assignment to treatment that such risk differentials become operative. Equally important, randomization should ensure that the distribution of other time origins (for example, onset of disease) is approximately the same across the treatment groups.

  This second criterion ordinarily overrides the first. In an experimental study of the effects of different kinds of marital counseling on the likelihood of divorce, the appropriate time origin would be the point at which couples were randomly assigned to treatment modality, *not* the date of the marriage. On the other hand, length of marriage at the time of treatment assignment can be included in the analysis as a covariate. This inclusion is *essential* if assignment to treatment was not randomized.

- *Choose the time origin that has the strongest effect on the hazard.* The main danger in choosing the *wrong* time origin is that the effect of time on the hazard may be inadequately controlled, leading to biased estimates of the effects of other covariates, especially time-dependent covariates. In general, the most important variables to control are those that have the biggest effects. For example, while it is certainly the case that the hazard for death is a function of age, the percent annual change in the hazard is rather small. On the other hand, the hazard for death due to ovarian cancer is likely to increase markedly from time since diagnosis. Hence, it is more important to control for time since diagnosis (by choosing it as the time origin). Again, it may be possible to control for other time origins by including them as covariates.

DATA STRUCTURE

The LIFETEST, LIFEREG, and PHREG procedures all expect data with the same basic structure. Indeed, this structure is fairly standard across many different computer packages for survival analysis. For each case in the sample, there must be one variable (which I'll call DUR) that contains either the time that an event occurred or, for censored cases, the last time at which that case was observed, both measured from the chosen origin. A second variable (which I'll call STATUS) is necessary if some of the cases are censored or if you want to distinguish different kinds of events. The STATUS variable is assigned arbitrary values that indicate the status of the individual at the time recorded in the DUR variable. If there is only one kind of event, it is common to have STATUS=1 for uncensored cases and STATUS=0 for censored cases, but any two values will do as long as you remember which is which. For PROC LIFEREG and PROC PHREG, which estimate regression models, the record should also contain values of the covariates. This is straightforward if the covariates are constant over time. The more complex data structure needed for time-dependent covariates is discussed in Chapter 5.

The basic data structure is illustrated by Output 2.1, which gives survival times for 25 patients diagnosed with myelomatosis (Peto et al., 1977). These patients were randomly assigned to two drug treatments, as indicated by the TREAT variable. The DUR variable gives the time in days from the point of randomization to either death or censoring (which could occur either by loss to follow up or termination of the observation). The STATUS variable has a value of 1 for those who died and 0 for those who

were censored. An additional covariate, RENAL, is an indicator variable
for normal (1) versus impaired (0) renal functioning at the time of
randomization. This data set is one of several that are analyzed in the
remaining chapters.

Output 2.1   *Myelomatosis Data*

OBS	DUR	STATUS	TREAT	RENAL
1	8	1	1	1
2	180	1	2	0
3	632	1	2	0
4	852	0	1	0
5	52	1	1	1
6	2240	0	2	0
7	220	1	1	0
8	63	1	1	1
9	195	1	2	0
10	76	1	2	0
11	70	1	2	0
12	8	1	1	0
13	13	1	2	1
14	1990	0	2	0
15	1976	0	1	0
16	18	1	2	1
17	700	1	2	0
18	1296	0	1	0
19	1460	0	1	0
20	210	1	2	0
21	63	1	1	1
22	1328	0	1	0
23	1296	1	2	0
24	365	0	1	0
25	23	1	2	1

Although the basic data structure for survival analysis is quite simple, it can often be an arduous task to get the data into this form, especially in complex life history studies that contain information on many different kinds of repeatable events. With its extremely flexible and powerful DATA step, SAS is well-suited to perform the kinds of programming necessary to process such complex data sets. Of particular utility is the rich set of date and time functions available in the DATA step. For example, suppose the origin time for some event is contained in three numeric variables: ORMONTH, ORDAY, and ORYEAR. Similarly, the event time is contained in the variables EVMONTH, EVDAY, and EVYEAR. To compute the number of days between origin and event time, you need only the statement

```
dur = MDY(evmonth,evday,evyear) - MDY(ormonth,orday,oryear);
```

The MDY function converts the month, day, and year into a SAS date: the number of days since January 1, 1960. Once that conversion takes place, simple subtraction suffices to get the duration in days. Many other functions are also available to convert time data in various formats into SAS date values.

CHAPTER **3**
Estimating and Comparing Survival Curves with PROC LIFETEST

INTRODUCTION

Prior to 1970, the estimation of survivor functions was the predominant method of survival analysis, and entire books were devoted to its exposition (for example, Gross and Clark, 1975). Nowadays, the workhorse of survival analysis is the Cox regression method discussed in Chapter 5, "Estimating Cox Regression Models with PROC PHREG." Nevertheless, survival curves are still useful for preliminary examination of the data, for computing derived quantities from regression models (like the median survival time or the 5-year probability of survival), and for evaluating the fit of regression models. For very simple experimental designs, standard tests for comparing survivor functions across treatment groups may suffice for analyzing the data. And in demography, the life-table method for estimating survivor functions still holds a preeminent place as a means of describing human mortality.

PROC LIFETEST produces estimates of survivor functions using either of two methods. The *Kaplan-Meier method* is more suitable when event times are measured with precision, especially if the number of observations is small. The *life-table* or *actuarial method* may be better for large data sets or when the measurement of event times is crude. In addition to computing and graphing the estimated survivor function, PROC LIFETEST provides several methods for testing the null hypothesis that survivor functions are identical for two or more groups (strata). Finally, PROC LIFETEST can test for associations between survival time and sets of quantitative covariates.

THE KAPLAN-MEIER METHOD

In biomedicine, the Kaplan-Meier (KM) estimator is the most widely used method for estimating survivor functions. Also known as the *product-limit estimator*, this method was known for many years prior to 1958 when Kaplan and Meier showed that it was, in fact, the nonparametric maximum likelihood estimator. This gave the method a solid theoretical justification.

When there are no censored data, the KM estimator is simple and intuitive. Recall from Chapter 2, "Basic Concepts of Survival Analysis," in the section **Describing Survival Distributions**, that the survivor function $S(t)$ is the probability that an event time is greater than t, where t can be any nonnegative number. When there is no censoring, the KM estimator $\hat{S}(t)$ is just the proportion of observations in the sample with event times greater than t. Thus, if 75 percent of the observations have event times greater than 5, we have $\hat{S}(5) = .75$.

The situation is also quite simple in the case of single right censoring (that is, when all the censored cases are censored at the same time c and all the observed event times are less than c). In that case, for all $t \leq c$, $\hat{S}(t)$ is still the sample proportion of observations with event times greater than t. For $t > c$, $\hat{S}(t)$ is undefined.

Things get more complicated when some censoring times are smaller than some event times. In that instance, the observed proportion of cases with event times greater than t can be biased downward because cases that are censored before t may, in fact, have "*died*" before t without our knowledge. The solution is as follows. Suppose there are k distinct event times, $t_1 < t_2 < \ldots < t_k$. At each time t_j, there are n_j individuals who are said to be at risk of an event. *At risk* means they have not experienced an event nor have they been censored prior to time t_j. If any cases are censored at exactly t_j, they are also considered to be at risk at t_j. Let d_j be the number of individuals who die at time t_j. The KM estimator is then defined as

$$\hat{S}(t) = \prod_{j:t_j \leq t}\left(1 - \frac{d_j}{n_j}\right) \qquad\qquad (3.1)$$

for $t_1 \leq t \leq t_k$. In words, this formula says that for a given time t, take all the event times that are less than or equal to t. For each of those event times, compute the quantity in brackets, which can be interpreted as an estimate of the conditional probability of surviving to time t_{j+1}, given that one has survived to time t_j. Then multiply all of these conditional probabilities together. For t less than t_1 (the smallest event time), $\hat{S}(t)$ is defined as 1.0. For t greater than t_k, the largest observed event time, the definition of $\hat{S}(t)$ depends on the configuration of the censored observations. When there are no censored times greater than t_k, $\hat{S}(t)$ is set to 0 for $t > t_k$. When there *are* censored times greater than t_k, $\hat{S}(t)$ is undefined for t greater than the largest censoring time. For an explanation of the rationale for equation (3.1), see **The Life-Table Method** later in this chapter.

Here's an example of how to get the KM estimator using PROC LIFETEST with the myelomatosis data shown in Output 2.1:

```
DATA myel;
   INPUT dur status treat renal;
   DATALINES;
   8         1         1         1
 180         1         2         0
 632         1         2         0
 852         0         1         0
  52         1         1         1
2240         0         2         0
 220         1         1         0
  63         1         1         1
 195         1         2         0
  76         1         2         0
  70         1         2         0
   8         1         1         0
  13         1         2         1
1990         0         2         0
1976         0         1         0
  18         1         2         1
 700         1         2         0
1296         0         1         0
1460         0         1         0
 210         1         2         0
  63         1         1         1
1328         0         1         0
1296         1         2         0
 365         0         1         0
  23         1         2         1
;
PROC LIFETEST DATA=myel;
   TIME dur*status(0);
RUN;
```

The KM estimator is the default, so you do not need to request it. If you want to be explicit, you can put METHOD=KM on the PROC LIFETEST statement. The syntax DUR*STATUS(0) is common to PROC LIFETEST, PROC LIFEREG, and PROC PHREG. The first variable is the time of the event or censoring, the second variable contains information on whether or not the observation was censored, and the numbers in parentheses (there can be more than one) are values of the second variable that correspond to censored observations. These statements produce the results shown in Output 3.1.

Output 3.1 *Kaplan-Meier Estimates for Myelomatosis Data*

```
                    Product-Limit Survival Estimates

                                    Survival
                                    Standard     Number    Number
       dur      Survival   Failure    Error      Failed     Left

       0.00      1.0000       0          0          0        25
       8.00        .          .          .          1        24
       8.00      0.9200     0.0800     0.0543       2        23
      13.00      0.8800     0.1200     0.0650       3        22
      18.00      0.8400     0.1600     0.0733       4        21
      23.00      0.8000     0.2000     0.0800       5        20
      52.00      0.7600     0.2400     0.0854       6        19
      63.00        .          .          .          7        18
      63.00      0.6800     0.3200     0.0933       8        17
      70.00      0.6400     0.3600     0.0960       9        16
      76.00      0.6000     0.4000     0.0980      10        15
     180.00      0.5600     0.4400     0.0993      11        14
     195.00      0.5200     0.4800     0.0999      12        13
     210.00      0.4800     0.5200     0.0999      13        12
     220.00      0.4400     0.5600     0.0993      14        11
     365.00*       .          .          .         14        10
     632.00      0.3960     0.6040     0.0986      15         9
     700.00      0.3520     0.6480     0.0970      16         8
     852.00*       .          .          .         16         7
    1296.00      0.3017     0.6983     0.0953      17         6
    1296.00*       .          .          .         17         5
    1328.00*       .          .          .         17         4
    1460.00*       .          .          .         17         3
    1976.00*       .          .          .         17         2
    1990.00*       .          .          .         17         1
    2240.00*       .          .          .         17         0

    NOTE: The marked survival times are censored observations.

           Summary Statistics for Time Variable dur

                     Quartile Estimates

                 Point           95% Confidence Interval
     Percent    Estimate    Transform    [Lower      Upper)

        75         .          LOGLOG      220.00        .
        50       210.00       LOGLOG       63.00     1296.00
        25        63.00       LOGLOG        8.00      180.00
```

(*continued*)

Output 3.1 *(continued)*

```
                    Mean      Standard Error

                  562.76           117.32

NOTE: The mean survival time and its standard error were underestimated
      because the largest observation was censored and the estimation was
      restricted to the largest event time.



        Summary of the Number of Censored and Uncensored Values

                                              Percent
                  Total   Failed    Censored  Censored

                    25      17          8       32.00
```

Each line of numbers in Output 3.1 corresponds to one of the 25 cases, arranged in ascending order (except for the first line, which is for time 0). Censored observations are starred. The crucial column is the second—labeled Survival—which gives the KM estimates. At 180 days, for example, the KM estimate is .56. We say, then, that the estimated probability that a patient will survive for 180 days or more is .56. When there are tied values (two or more cases that die at the same reported time), as we have at 8 days and 63 days, the KM estimate is reported only for the last of the tied cases. No KM estimates are reported for the censored times.

In fact, however, the KM estimator is defined for any time between 0 and the largest event or censoring time. It's just that it changes only at an observed event time. Thus, the estimated survival probability for any time from 70 days up to (but not including) 76 days is .64. The 1-year (365 days) survival probability is .44, the same as it was at 220 days. After 2,240 days (the largest censoring time), the KM estimate is undefined.

The third column, labeled Failure, is just 1 minus the KM estimate, which is the estimated probability of a death prior to the specified time. Thus, it is an estimate of the cumulative distribution function. The fourth column, labeled Survival Standard Error, is an estimate of the standard error of the KM estimate, obtained by the well-known Greenwood's formula (Collett, 2003). Although this standard error could be used directly to construct confidence intervals around the survival probabilities, PROC LIFETEST has better methods that we shall examine later.

The fifth column, labeled Number Failed, is just the cumulative number of cases that experienced events prior to and including each point in time. The column labeled Number Left is the number of cases that have neither experienced events nor been censored prior to each point in time. This is the size of the *risk set* for each time point. Below the main table, you find the estimated 75th, 50th, and 25th percentiles (labeled Quartile Estimates). If these were not given, you could easily determine them from the Failure column. Thus, the 25th percentile (63 in this case) is the smallest event time such that the probability of dying earlier is greater than .25. No value is reported for the 75th percentile because the KM estimator for these data never reaches a failure probability greater than .70.

Of greatest interest is the 50th percentile, which is, of course, the median death time. Here, the median is 210 days, with a 95% confidence interval of 63 to 1296. As noted in the table, the confidence intervals are calculated using a log-log transform that preserves the upper bound of 1 and the lower bound of 0 on the survival probabilities. Although other transforms are optionally available, there is rarely any need to use them.

An estimated mean time of death is also reported. This value is calculated directly from the estimated survival function. It is not simply the mean time of death for those who died. As noted in the output, the mean is biased downward when there are censoring times greater than the largest event time. Even when this is not the case, the upper tail of the distribution will be poorly estimated when a substantial number of the cases are censored, and this can greatly affect estimates of the mean. Consequently, the median is typically a better measure of central tendency for censored survival data.

Usually you will want to see a plot of the estimated survival function. PROC LIFETEST can produce plots in three different modes: line printer, traditional graphics, and the output delivery system (ODS). Here I use ODS because it has a much richer array of features. Here's how to get the survivor function graph using ODS:

```
ODS GRAPHICS ON;
PROC LIFETEST DATA=myel PLOTS=S;
   TIME dur*status(0);
RUN;
ODS GRAPHICS OFF;
```

The resulting graph (shown in Output 3.2) can be accessed by double-clicking on the graph icon in the Results window.

Output 3.2   *Plot of the Survivor Function for Myelomatosis Data*

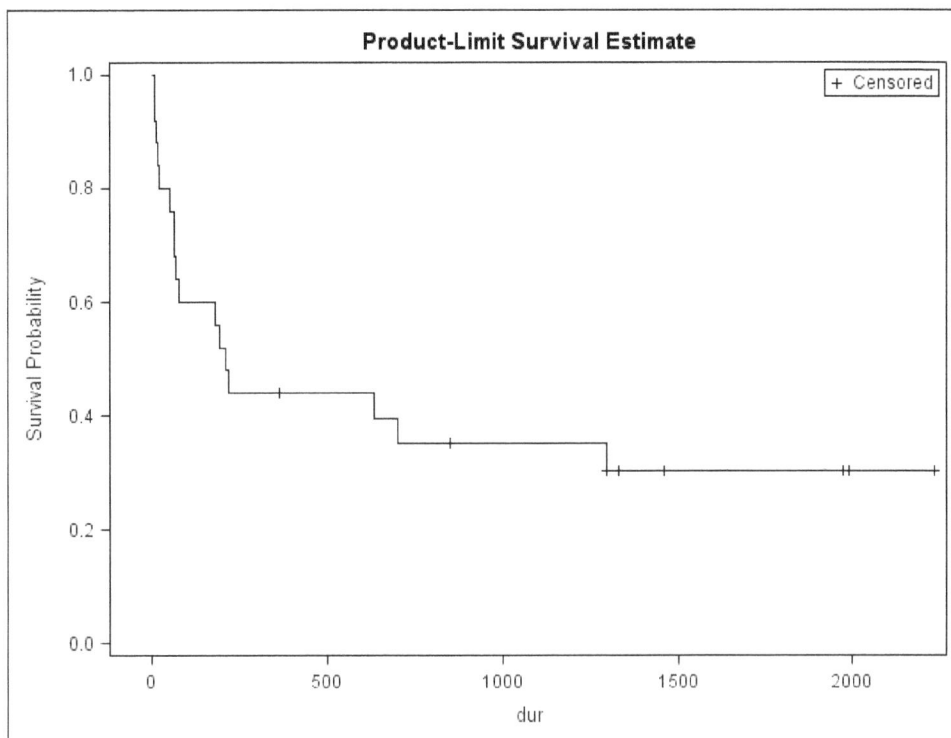

Notice that the graph contains plus signs (that look like tick marks) wherever there are censored observations. These marks can be very distracting when the data set is large with lots of censored observations. They can be suppressed by using the NOCENSOR option. Another useful option is ATRISK, which adds the number of individuals still at risk (not yet dead or censored) to the graph. To get a graph with 95% confidence limits around the survivor function, use the CL option. All three options can be specified with the statement

```
PROC LIFETEST DATA=myel PLOTS=S(NOCENSOR ATRISK CL);
```

which produces the graph shown in Output 3.3.

Output 3.3   *Plot of the Survivor Function for Myelomatosis Data with Pointwise Confidence Limits*

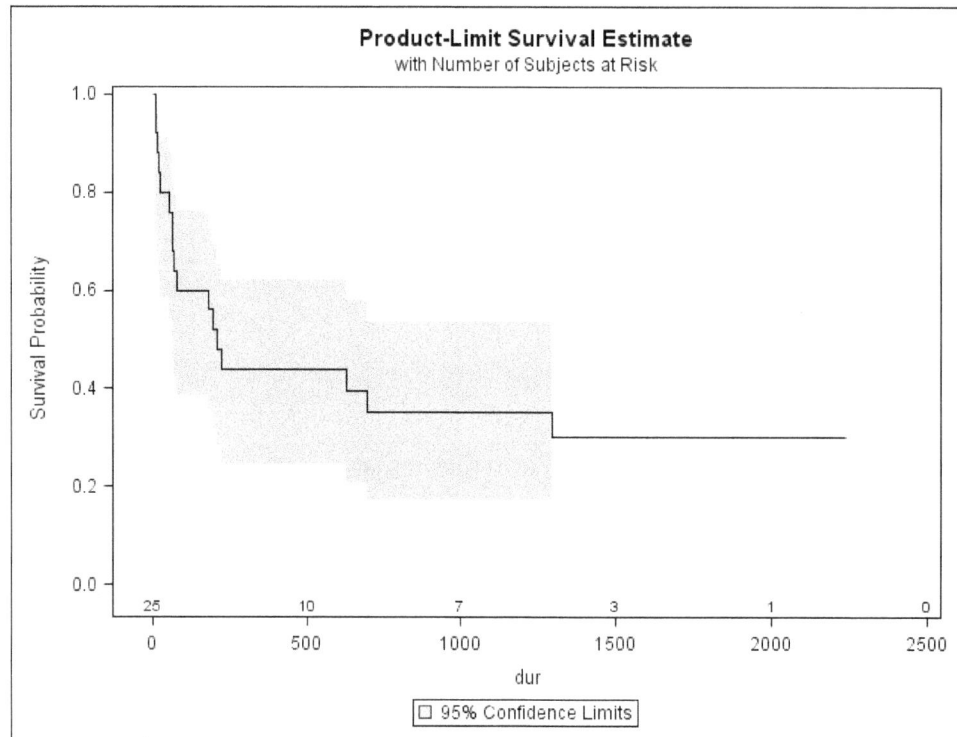

The confidence limits shown in Output 3.3 are *pointwise limits*, meaning that for each specified survival time, we are 95% confident that the probability of surviving to that time is within those limits. Note that the confidence limits only extend to the largest event time.

Suppose we want confidence *bands* that can be interpreted by saying that we are 95% confident that the entire survivor function falls within the upper curve and the lower curve. More complex methods are needed to produce such bands, and PROC LIFETEST offers two: the Hall-Wellner method and the equal precision (EP) method. I prefer the EP method because it tends to produce confidence bands that are more stable in the tails. To implement this method, the option becomes PLOTS=S(CB=EP). To get both pointwise and EP bands, use the option PLOTS=S(CL CB=EP), which produces Output 3.4. As is evident in that graph, the confidence bands are always wider than the pointwise confidence limits.

Output 3.4   *Survivor Function for Myelomatosis Data with Confidence Bands and Pointwise Limits*

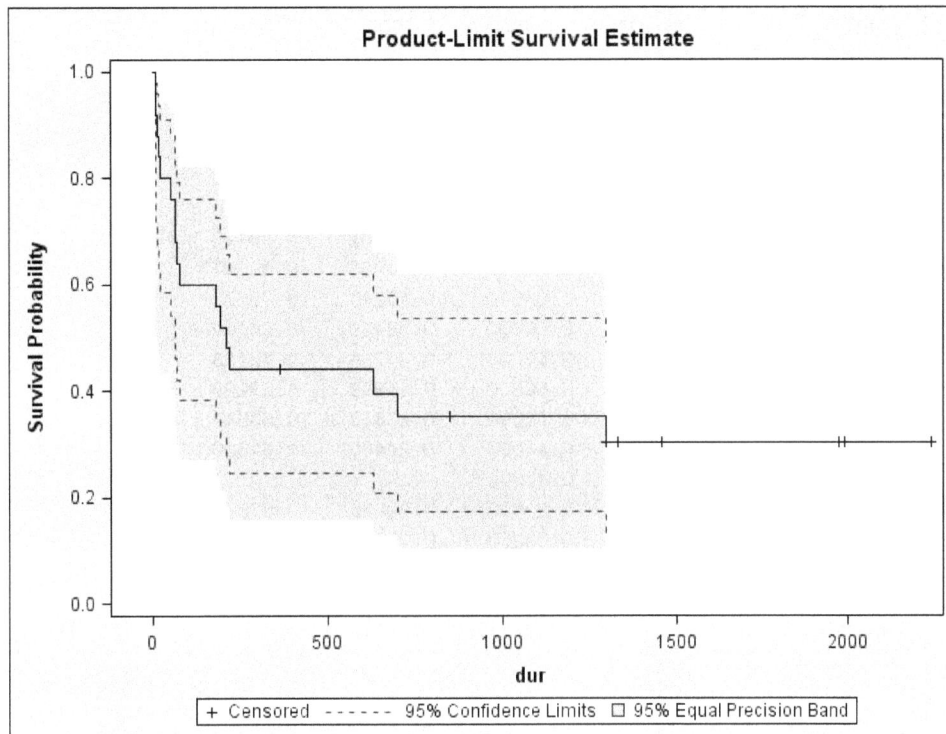

You can also write the pointwise limits to a SAS data set using the OUTSURV option:

```
PROC LIFETEST DATA=myel OUTSURV=a;
   TIME dur*status(0);
PROC PRINT DATA=a;
RUN;
```

Output 3.5 shows the printed data set. Note that the data set contains one record for each unique event or censoring time, plus a record for time 0. The two right-hand columns give the upper and lower limits for a 95% confidence interval around each survival probability. (To get 90% intervals, use ALPHA=.10 as an option in the PROC LIFETEST statement.)

Output 3.5 *Data Set Produced by the OUTSURV= Option*

Obs	dur	_CENSOR_	SURVIVAL	SDF_LCL	SDF_UCL
1	0	.	1.00000	1.00000	1.00000
2	8	0	0.92000	0.71639	0.97937
3	13	0	0.88000	0.67256	0.95964
4	18	0	0.84000	0.62806	0.93673
5	23	0	0.80000	0.58445	0.91146
6	52	0	0.76000	0.54205	0.88428
7	63	0	0.68000	0.46093	0.82527
8	70	0	0.64000	0.42215	0.79378
9	76	0	0.60000	0.38449	0.76109
10	180	0	0.56000	0.34794	0.72728
11	195	0	0.52000	0.31249	0.69238
12	210	0	0.48000	0.27813	0.65640
13	220	0	0.44000	0.24490	0.61936
14	365	1	0.44000	.	.
15	632	0	0.39600	0.20826	0.57872
16	700	0	0.35200	0.17355	0.53660
17	852	1	0.35200	.	.
18	1296	0	0.30171	0.13419	0.48925
19	1296	1	.	.	.
20	1328	1	.	.	.
21	1460	1	.	.	.
22	1976	1	.	.	.
23	1990	1	.	.	.
24	2240	1	.	.	.

By default, these limits are calculated by adding and subtracting 1.96 standard errors to the log-log transformation of the survivor function, $\log(-\log \hat{S}(t))$, and then reversing the transformation to get back to the original metric. This method ensures that the confidence limits will not be greater than 1 or less than 0. Other transformations are available, the most attractive being the logit: $\log[\hat{S}(t)/(1-\hat{S}(t))]$. To switch to this transform, include the CONFTYPE=LOGIT option in the PROC statement. You can also write the equal precision confidence bands to the OUTSURV data set by including the CONFBAND=EP option in the PROC statement.

TESTING FOR DIFFERENCES IN SURVIVOR FUNCTIONS

If an experimental treatment has been applied to one group but not another, the obvious question to ask is "Did the treatment make a difference in the survival experience of the two groups?" Because the survivor function gives a complete accounting of the survival experience of each group, a natural approach to answering this question is to test the null hypothesis that the survivor functions are the same in the two groups (that is, $S_1(t) = S_2(t)$ for all t, where the subscripts distinguish the two groups). PROC LIFETEST can calculate several different statistics for testing this null hypothesis.

For the myelomatosis data, it is clearly desirable to test whether the treatment variable (TREAT) has any effect on the survival experience of the two groups. To do that with PROC LIFETEST, simply add a STRATA statement after the TIME statement:

```
ODS GRAPHICS ON;
PROC LIFETEST DATA=myel PLOTS=S(TEST);
   TIME dur*status(0);
   STRATA treat;
RUN;
ODS GRAPHICS OFF;
```

The STRATA statement has three consequences:

1. First, instead of a single table with KM estimates, separate tables (not shown here) are produced for each of the two treatment groups.
2. Second, corresponding to the two tables are two graphs of the survivor function, superimposed on the same axes for easy comparison.
3. Third, PROC LIFETEST reports several statistics related to testing for differences between the two groups. Also, the TEST option (after PLOTS=S) includes the log-rank test in the survivor plot.

Look at the graphs in Output 3.6. Before 220 days, the two survival curves are virtually indistinguishable, with little visual evidence of a treatment effect. The gap that develops after 220 days reflects the fact that no additional deaths occur in treatment group 1 after that time.

Output 3.6 *Survival Curves for Two Treatment Groups*

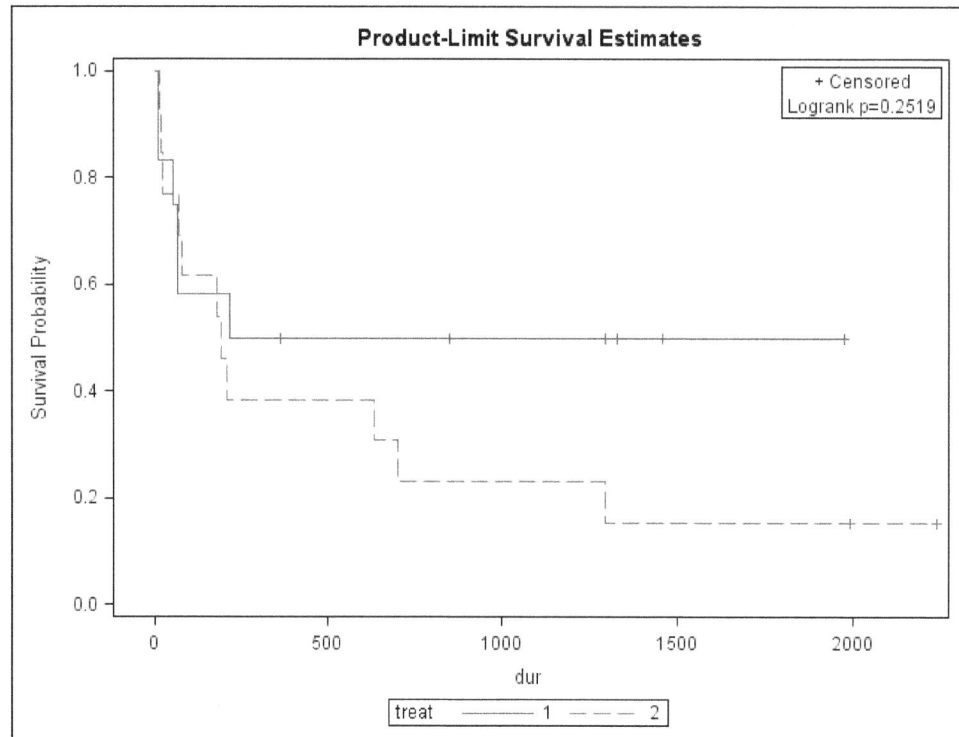

Next, PROC LIFETEST prints log-rank and Wilcoxon statistics for each treatment group, followed by an estimate of their covariance matrix. These are used to compute the chi-square statistics shown near the bottom of Output 3.7. In most cases, you can ignore the preliminaries and look only at the *p*-values. Here they are far from significant, which is hardly surprising given the graphical results and the small sample size. Thus, there is no evidence that would support the rejection of the null hypothesis that the two groups have exactly the same survivor function (that is, exactly the same probability distribution for the event times). The *p*-value for a likelihood-ratio test (-2log(LR)) is also reported; this test is usually inferior to the other two because it requires the unnecessary assumption that the hazard function is constant in each group, implying an exponential distribution for event times.

Output 3.7 *Statistics for Testing for Differences between Two Groups*

```
           Testing Homogeneity of Survival Curves for dur over Strata

                              Rank Statistics

              treat        Log-Rank      Wilcoxon

                1           -2.3376       -18.000
                2            2.3376        18.000

           Covariance Matrix for the Log-Rank Statistics

              treat               1              2

                1             4.16301       -4.16301
                2            -4.16301        4.16301

           Covariance Matrix for the Wilcoxon Statistics

              treat               1              2

                1             1301.00       -1301.00
                2            -1301.00        1301.00

                      Test of Equality over Strata

                                              Pr >
              Test        Chi-Square      DF    Chi-Square

              Log-Rank      1.3126        1      0.2519
              Wilcoxon      0.2490        1      0.6178
              -2Log(LR)     1.5240        1      0.2170
```

The log-rank test is the best known and most widely used test for differences in the survivor function, but the Wilcoxon test is also popular. Are there any reasons for choosing one over the other? Each statistic can be written as a function of deviations of observed numbers of events from expected numbers. For group 1, the log-rank statistic can be written as

$$\sum_{j=1}^{r}(d_{1j} - e_{1j})$$

where the summation is over all unique event times (in both groups), and there are a total of r such times. d_{1j} is the number of deaths that occur in group 1 at time j, and e_{1j} is the expected number of events in group 1 at time j. The expected number is given by $n_{1j}d_j/n_j$, where n_j is the total number of cases that are at risk just prior to time j, n_{1j} is the number at risk just prior to time j in group 1, and d_j is the total number of deaths at time j in both groups. (This is just the usual formula for computing expected cell counts in a 2×2 table, under the hypothesis of independence.) As shown in Output 3.7, the log-rank statistic in group 1 is −2.3376. The chi-square statistic is calculated by squaring this number and dividing by the estimated variance, which is 4.16301 in this case.

The Wilcoxon statistic, given by

$$\sum_{j=1}^{r} n_j(d_{1j} - e_{1j})$$

differs from the log-rank statistic only by the presence of n_j, the total number at risk at each time point. Thus, it is a *weighted* sum of the deviations of observed numbers of events from expected numbers of events. As with the log-rank statistic, the chi-square test is calculated by squaring the Wilcoxon statistic for either group (−18 for group 1 in this example) and dividing by the estimated variance (1301).

Because the Wilcoxon test gives more weight to early times than to late times (n_j never increases with time), it is less sensitive than the log-rank test to differences between groups that occur at later points in time. To put it another way, although both statistics test the same null hypothesis, they differ in their sensitivity to various kinds of departures from that hypothesis. In particular, the log-rank test is more powerful for detecting differences of the form

$$S_1(t) = [S_2(t)]^\gamma$$

where γ is some positive number other than 1.0. This equation defines a *proportional hazards model,* which is discussed in detail in Chapter 5. (As we will see in that chapter, the log-rank test is closely related to tests for differences between two groups that are performed within the framework of Cox's proportional hazards model.) In contrast, the Wilcoxon test is more powerful than the log-rank test in situations where event times have log-normal distributions (discussed in the next chapter) with a common variance but with different means in the two groups. Neither test is particularly good at detecting differences when survival curves cross.

Now let's see an example where the survival distributions clearly differ across two groups. Using the same data set, we can stratify on RENAL, the variable that has a value of 1 if renal functioning was impaired at the time of randomization; otherwise, the variable has a value of 0. Output 3.8 shows that the survival curve for those with impaired renal functioning drops precipitously, while the curve for those with normal functioning declines much more gradually. The three hypothesis tests are unanimous in rejecting the null hypothesis of no difference between the two groups.

Output 3.8 *Graphs and Tests for Stratifying by Renal Functioning*

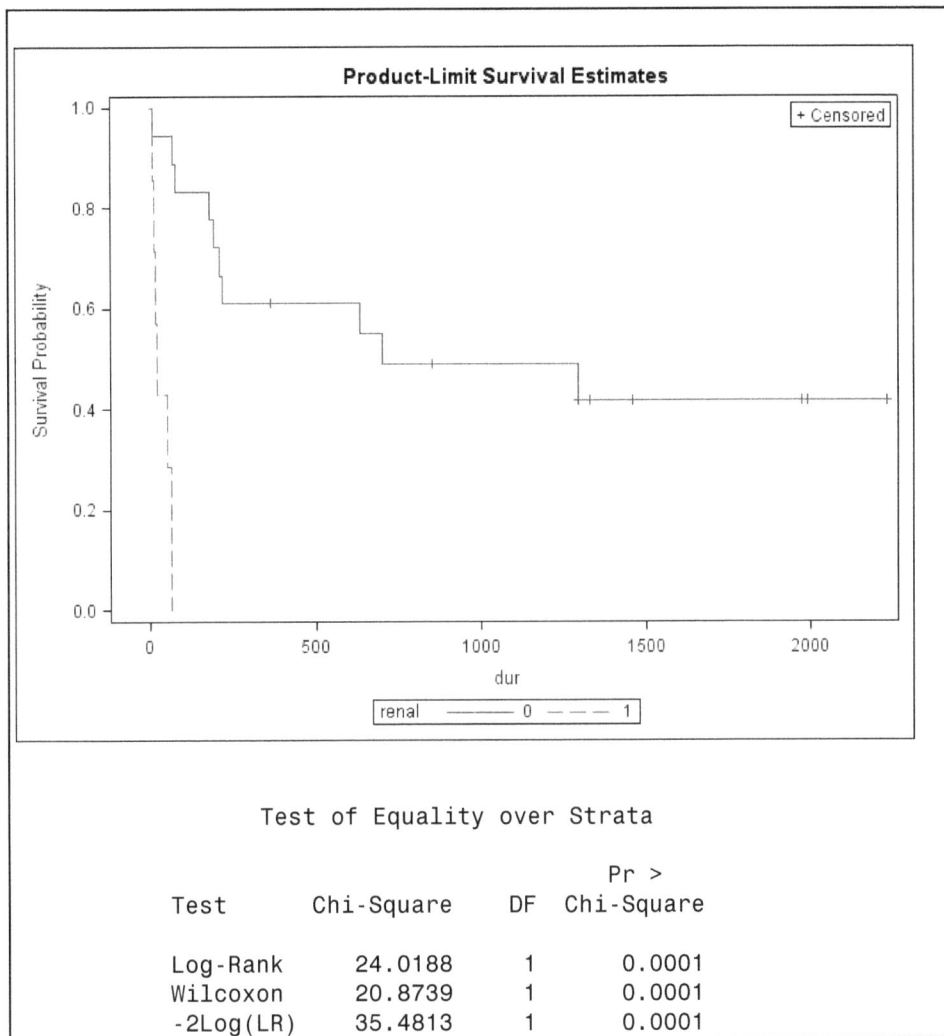

Product-Limit Survival Estimates

Test of Equality over Strata

Test	Chi-Square	DF	Pr > Chi-Square
Log-Rank	24.0188	1	0.0001
Wilcoxon	20.8739	1	0.0001
-2Log(LR)	35.4813	1	0.0001

In addition to the log-rank and Wilcoxon tests, which are produced by default, the STRATA statement also has options for four other nonparametric tests: Tarone-Ware, Peto-Peto, modified Peto-Peto, and Fleming-Harrington. Like the Wilcoxon and log-rank tests, all of these can be represented as a weighted sum of observed and expected numbers of events:

$$\sum_{j=1}^{r} w_j (d_{1j} - e_{1j})$$

For the Tarone-Ware test, w_j is the square root of n_j, so this test behaves much like the Wilcoxon test in being more sensitive to differences at earlier rather than later times. That's also true of the two Peto tests, for which w_j is a function of the survivor function itself. Fleming-Harrington is actually a family of tests in which the weights depend on two parameters, p and q, which can be chosen by the user:

$$w_j = \hat{S}(t_j)^p [1 - \hat{S}(t_j)]^q$$

Although p and q can be any nonnegative numbers, here are a few special cases. When both p and q are 0, you get the log-rank test. When p is 1 and q is 0, you get something very close to the Peto-Peto test. When q is 1 and p is 0, w_j increases with time, unlike any of the other tests. When both q and p are 1, you get a weight function that is maximized at the median and gets small for large or small times.

To get all these tests, the STRATA statement can be changed to

```
STRATA treat / TESTS=ALL;
```

which produces the results in Output 3.9. By default, the Fleming-Harrington test is performed with $p=1$ and $q=0$.

Output 3.9 *Optional Tests for Differences between Survival Functions*

```
                    Test of Equality over Strata

                                                  Pr >
              Test           Chi-Square    DF    Chi-Square

              Log-Rank         1.3126       1      0.2519
              Wilcoxon         0.2490       1      0.6178
              Tarone           0.6514       1      0.4196
              Peto             0.3766       1      0.5394
              Modified Peto    0.3321       1      0.5644
              Fleming(1)       0.3045       1      0.5811
```

So far, we have only tested for differences between two groups. The tests readily generalize to three or more groups, with the null hypothesis that all groups have the same survivor function. If the null hypothesis is true, the test statistics all have chi-square distributions with degrees of freedom equal to the number of groups minus 1. The STRATA statement provides great flexibility in defining those groups. PROC LIFETEST constructs a stratum corresponding to each unique value of the variable in the STRATA statement. That variable may be either numeric or character. If the STRATA statement contains two or more variable names, PROC LIFETEST constructs one group for every unique combination of values of the variables.

To illustrate this and other options, we need a different data set. Output 3.10 displays the first 20 observations from the recidivism study that was briefly described in Chapter 1, "Introduction," in the section **Why Use Survival Analysis?** The sample consisted of 432 male inmates who were released from Maryland state prisons in the early 1970s (Rossi et al., 1980). These men were followed for 1 year after their release, and the dates of any arrests were recorded. We'll look only at the first arrest here. (In fact, there weren't enough men with two or more arrests to use the techniques for repeated events discussed in Chapter 8, "Heterogeneity, Repeated Events, and Other Topics.") The WEEK variable contains the week of the first arrest after release. The variable ARREST has a value of 1 for those who were arrested during the 1-year follow-up, and it has a value of 0 for those who were not. Only 26 percent of the men were arrested. The data are singly right censored so that all the censored cases have a value of 52 for WEEK.

Output 3.10   *Recidivism Data (First 20 Cases Out of 432)*

OBS	WEEK	ARREST	FIN	AGE	RACE	WEXP	MAR	PARO	PRIO
1	52	0	1	24	1	1	0	1	1
2	52	0	0	29	1	1	0	1	3
3	52	0	1	20	1	1	1	1	1
4	52	0	0	20	1	0	0	1	1
5	52	0	1	31	0	1	0	1	3
6	12	1	1	22	1	1	1	1	2
7	52	0	0	24	1	1	0	1	2
8	19	1	0	18	1	0	0	0	2
9	52	0	0	18	1	0	0	1	3
10	15	1	1	22	1	0	0	1	3
11	8	1	1	21	1	1	0	1	4
12	52	0	1	21	1	0	0	1	1
13	52	0	1	21	0	1	0	1	1
14	36	1	1	19	1	0	0	1	2
15	52	1	0	33	1	1	0	1	2
16	4	1	0	18	1	1	0	0	1
17	45	1	1	18	1	0	0	0	5
18	52	1	0	21	1	0	0	0	0
19	52	0	1	20	1	0	1	0	1
20	52	0	0	22	1	1	0	0	1

The covariates shown in Output 3.10 are defined as follows:

FIN   has a value of 1 if the inmate received financial aid after release; otherwise, FIN has a value of 0. This variable was randomly assigned with equal numbers in each category.

AGE   specifies age in years at the time of release.

RACE   has a value of 1 if the person was black; otherwise, RACE has a value of 0.

WEXP   has a value of 1 if the inmate had full-time work experience before incarceration; otherwise, WEXP has a value of 0.

MAR   has a value of 1 if the inmate was married at the time of release; otherwise, MAR has a value of 0.

PARO   has a value of 1 if the inmate was released on parole; otherwise, PARO has a value of 0.

PRIO   specifies the number of convictions an inmate had prior to incarceration.

The following program stratifies by the four combinations of WEXP and PARO. The ADJUST option (new in SAS 9.2) tells PROC LIFETEST to produce *p*-values for all six pairwise comparisons of the four strata and then to report *p*-values that have been adjusted for multiple comparisons using Tukey's method (other methods are also available):

```
PROC LIFETEST DAta=my.recid;
  TIME week*arrest(0);
  STRATA wexp paro / ADJUST=TUKEY;
RUN;
```

Results are shown in Output 3.11. The first table shows how the four strata correspond to the possible combinations of WEXP and PARO. Next we get overall chi-square tests of the null hypothesis that the survivor functions are identical across the four strata. All three tests are significant at the .05 level. Last, we see the log-rank tests comparing each possible pair of strata. Three of the tests are significant using the "raw" *p*-value, but only one of these comparisons is significant after the Tukey adjustment. (Not shown is a similar table for the Wilcoxon test.)

Output 3.11 *Tests Comparing More Than Two Groups, with Multiple Comparison Adjustment*

Stratum	wexp	paro
1	0	0
2	0	1
3	1	0
4	1	1

Test of Equality over Strata

Test	Chi-Square	DF	Pr > Chi-Square
Log-Rank	10.2074	3	0.0169
Wilcoxon	11.2596	3	0.0104
-2Log(LR)	9.2725	3	0.0259

Adjustment for Multiple Comparisons for the Logrank Test

			p-Values	
Stratum	_Stratum_	Chi-Square	Raw	Tukey-Kramer
1	2	0.00542	0.9413	0.9999
1	3	4.5853	0.0322	0.1401
1	4	6.6370	0.0100	0.0491
2	3	3.5427	0.0598	0.2357
2	4	5.2322	0.0222	0.1009
3	4	0.4506	0.5021	0.9080

For numeric variables, you can use the STRATA statement to define groups by intervals rather than by unique values. For the recidivism data, the AGE variable can be stratified as

```
STRATA age(21 24 28);
```

This statement produces four strata, corresponding to the intervals:

$$age < 21$$
$$21 \leq age < 24$$
$$24 \leq age < 28$$
$$28 \leq age$$

Output 3.12 displays the results. Here I have also requested multiple comparisons but now using the Bonferroni adjustment (ADJUST=BON). Note that except for the two extreme strata, each stratum is identified by the midpoint of the interval. Clearly one must reject the null hypothesis that all age strata have the same survivor function. Three of the six pairwise comparisons are also significant, either with or without the adjustment.

Output 3.12   *Tests for Multiple Age Groups*

```
       Summary of the Number of Censored and Uncensored Values


                                                            Percent
    Stratum    age          Total   Failed    Censored    Censored


        1      <21           127      52          75        59.06
        2       22.5         114      29          85        74.56
        3       26            91      17          74        81.32
        4      >=28          100      16          84        84.00
    --------------------------------------------------------------
     Total                   432     114         318        73.61



              Test of Equality over Strata


                                                Pr >
              Test      Chi-Square      DF    Chi-Square


              Log-Rank    22.2316        3      <.0001
              Wilcoxon    20.9222        3      0.0001
              -2Log(LR)   19.8330        3      0.0002



     Adjustment for Multiple Comparisons for the Logrank Test
        Strata Comparison                        p-Values
           age          age     Chi-Square     Raw     Bonferroni


       21.0000       22.5000        8.4518    0.0036      0.0219
       21.0000            26       15.6465    <.0001      0.0005
       21.0000       28.0000       18.3765    <.0001      0.0001
       22.5000            26        0.8272    0.3631      1.0000
       22.5000       28.0000        1.6912    0.1934      1.0000
            26       28.0000        0.1836    0.6683      1.0000
```

THE LIFE-TABLE METHOD

If the number of observations is large and if event times are precisely measured, there will be many unique event times. The KM method then produces long tables that may be unwieldy for presentation and interpretation. One way to solve this problem is to use the TIMELIST option (in the PROC LIFEREG statement), which reports the KM estimates only at specified points in time. An alternative solution is to switch to the life-table method, in which event times are grouped into intervals that can be as long or short as you please. In addition, the life-table method (also known as the *actuarial method*) can produce estimates and plots of the hazard function. The downside to the life-table method is that the choice of intervals is usually somewhat arbitrary, leading to arbitrariness in the results and possible uncertainty about how to choose the intervals. There is inevitably some loss of information as well. Note, however, that PROC LIFETEST computes the log-rank and Wilcoxon statistics (as well as other optional test statistics) from the *ungrouped* data (if available) so they are unaffected by the choice of intervals for the life-table method.

We now request a life table for the recidivism data, using the default specification for interval lengths:

```
PROC LIFETEST DATA=recid METHOD=LIFE;
   TIME week*arrest(0);
RUN;
```

Output 3.13 shows the results. PROC LIFETEST constructs six intervals, starting at 0 and incrementing by periods of 10 weeks. The algorithm for the default choice of intervals is fairly complex (see the *SAS/STAT User's Guide* for details). You can override the default by specifying WIDTH=w in the PROC LIFETEST statement. The intervals will then begin with [0, w) and will increment by w. Alternatively, you can get even more control over the intervals by specifying INTERVALS=a b c ... in the PROC LIFETEST statement, where a, b, and c are cut points. For example, INTERVALS= 15 20 30 50 produces the intervals [0, 15), [15, 20), [20, 30), [30, 50), [50, ∞). See the *SAS/STAT User's Guide* for other options. Note that intervals do not have to be the same length. It's often desirable to make later intervals longer so that they include enough events to get reliable estimates of the hazard and other statistics.

Output 3.13 *Applying the Life-Table Method to the Recidivism Data*

```
                          Life Table Survival Estimates


                                                       Conditional
                                      Effective  Conditional  Probability
        Interval      Number   Number   Sample    Probability   Standard
  [Lower,     Upper)   Failed  Censored   Size     of Failure     Error

       0         10      14        0     432.0       0.0324      0.00852
      10         20      21        0     418.0       0.0502      0.0107
      20         30      23        0     397.0       0.0579      0.0117
      30         40      23        0     374.0       0.0615      0.0124
      40         50      26        0     351.0       0.0741      0.0140
      50         60       7      318     166.0       0.0422      0.0156

                                         Survival     Median      Median
        Interval                         Standard    Residual    Standard
  [Lower,     Upper)  Survival  Failure    Error     Lifetime     Error

       0         10    1.0000       0         0          .          .
      10         20    0.9676    0.0324    0.00852       .          .
      20         30    0.9190    0.0810    0.0131        .          .
      30         40    0.8657    0.1343    0.0164        .          .
      40         50    0.8125    0.1875    0.0188        .          .
      50         60    0.7523    0.2477    0.0208        .          .

                  Evaluated at the Midpoint of the Interval

                                    PDF                   Hazard
        Interval                 Standard               Standard
  [Lower,     Upper)    PDF        Error      Hazard      Error

       0         10   0.00324    0.000852   0.003294    0.00088
      10         20   0.00486    0.00103    0.005153    0.001124
      20         30   0.00532    0.00108    0.005966    0.001244
      30         40   0.00532    0.00108    0.006345    0.001322
      40         50   0.00602    0.00114    0.007692    0.001507
      50         60   0.00317    0.00118    0.004308    0.001628


        Summary of the Number of Censored and Uncensored Values


                                            Percent
                Total  Failed    Censored   Censored


                 432     114        318      73.61
```

For each interval, 14 different statistics are reported. The four statistics displayed in the first panel, while not of major interest in themselves, are necessary for calculating the later statistics. Number Failed and Number Censored should be self-explanatory. Effective Sample Size is straightforward for the first five intervals because they contain no censored cases. The effective sample size for these intervals is just the number of persons who had not yet been arrested at the start of the interval. For the last interval, however, the effective sample size is only 166, even though 351 persons made it to the 50th week without an arrest. Why? The answer is a fundamental property of the life-table method. The method treats any cases censored within an interval as if they were censored *at the midpoint of the interval.* This treatment is equivalent to assuming that the distribution of censoring times is uniform within the interval. Because censored cases are only at risk for half of the interval, they only count for half in figuring the effective sample size. Thus, the effective sample size for the last interval is 7+(318/2)=166. The 7 corresponds to the seven men who *were* arrested in the interval; they are treated as though they were at risk for the whole interval.

The Conditional Probability of Failure is an estimate of the probability that a person will be arrested in the interval, given that he made it to the start of the interval. This estimate is calculated as (number failed)/(effective sample size). An estimate of its standard error is given in the next panel. The Survival column is the life-table estimate of the survivor function (that is, the probability that the event occurs at a time greater than or equal to the start time of each interval). For example, the estimated probability that an inmate will not be arrested until week 30 or later is .8657.

The survival estimate is calculated from the conditional probabilities of failure in the following way. For interval i, let t_i be the start time and q_i be the estimated conditional probability of failure. The probability of surviving to t_i or beyond is then

$$\hat{S}(t) = \prod_{j=1}^{i-1} (1 - q_j).$$ (3.2)

For i = 1 and, hence, t_i = 0, the survival probability is set to 1.0.

The rationale for equation (3.2) is a fairly simple application of conditional probability theory. Suppose we want an expression for the probability of surviving to t_4 or beyond. To obtain this, let's define the following events:

    A = survival to t_2 or beyond.
    B = survival to t_3 or beyond.
    C = survival to t_4 or beyond.

We want the probability of C. But since you can't get past t_4 without getting past t_2 and t_3, we can write $\Pr(C) = \Pr(A, B, C)$. By the definition of conditional probability, we can rewrite this as

$$\Pr(A, B, C)=\Pr(C\mid A, B)\Pr(B\mid A)\Pr(A)=(1 - q_3)(1 - q_2)(1 - q_1).$$

Extending this argument to other intervals yields the formula in equation (3.2). Note the similarity between this formula and equation (3.1) for the KM estimator. In equation (3.1), d_j/n_j is equivalent to q_j in equation (3.2); both are estimates of the probability of failure in an interval given survival to the start of the interval. The major differences between the two formulas are as follows:

- The number of censored observations in an interval is not halved in the KM estimator.
- The interval boundaries for the KM estimator are determined by the event times themselves.

Thus, each interval for KM estimation extends from one unique event time up to, but not including, the next unique event time.

Continuing with the second panel of Output 3.13, the Failure column is just 1 minus the Survival column. We are also given the standard errors of the Survival probabilities. The Median Residual Lifetime column is, in principle, an estimate of the remaining time until an event for an individual who survived to the start of the interval. For this example, however, the estimates are all missing. To calculate this statistic for a given interval, there must be a later interval whose survival probability is less than half the survival probability associated with the interval of interest. It is apparent from Output 3.13 that no interval satisfies this criterion. For many data sets, there will be at least some later intervals for which this statistic cannot be calculated.

The PDF Standard Error column gives the estimated value of the probability density function at the midpoint of the interval. Of greater interest is the Hazard column, which gives estimates of the hazard function evaluated at the midpoint of each interval. This is calculated as

(3.3)

$$h(t_{im}) = \frac{d_i}{b_i\left(n_i - \frac{w_i}{2} - \frac{d_i}{2}\right)}$$

where for the *i*th interval, t_{im} is the midpoint, d_i is the number of events, b_i is the width of the interval, n_i is the number still at risk at the beginning of the interval, and w_i is the number of cases withdrawn (censored) within the interval. A better estimate of the hazard could be obtained by d_i/T_i, where T_i is the total exposure time within interval *i*. For each individual, exposure time is the amount of time actually observed within the interval. For an individual who is known to have survived the whole interval, exposure time is just the interval width b_j. For individuals who had events or who withdrew in the interval, exposure time is the time from the beginning of the interval until the event or withdrawal. *Total exposure time* is the sum of all the individual exposure times. The denominator in equation (3.3) is an approximation to total exposure time, such that all events and all withdrawals are presumed to occur at the midpoint of the interval (thus, the division by 2). Why use an inferior estimator? Well, exact exposure times are not always available (see the next section), so the estimator in equation (3.3) has become the standard for life tables.

You can get plots of the survival and hazard estimates by using PLOTS=(S,H) in the PROC LIFETEST statement, as in the following program:

```
ODS GRAPHICS ON;
PROC LIFETEST DATA=recid METHOD=LIFE PLOTS=(S,H);
   TIME week*arrest(0);
RUN;
ODS GRAPHICS OFF;
```

Output 3.14 displays the graph of the hazard function. Apparently, the hazard of arrest increases steadily until the 50-60 week interval, when it drops precipitously from .077 to .043. This drop is an artifact of the way that the last interval is constructed, however. Although the interval runs from 50 to 60, in fact, no one was at risk of an arrest after week 52 when the study was terminated. As a result, the denominator in equation (3.3) is a gross overestimate of the exposure time in the interval.

Output 3.14 *Hazard Estimates for Recidivism Data*

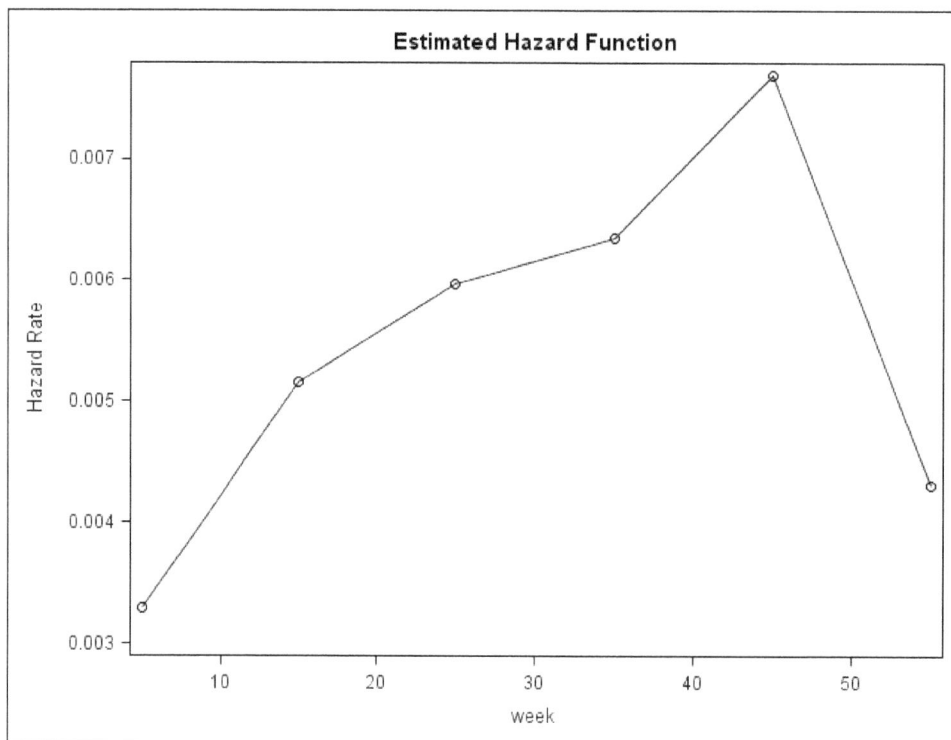

We can fix this by explicitly setting the last interval to end at 53. (If we set it at 52, the interval will not include arrests that occurred in week 52 because the interval is open on the right). At the same time, it is better to recode the censored times from 52 to 53 because they are not censored *within* the interval but rather at the end. Recoding effectively credits the full interval (rather than only half) as exposure time for the censored cases.

Here's the revised code:

```
DATA newrecid;
   SET recid;
   IF arrest=0 THEN week=53;
PROC LIFETEST DATA=newrecid METHOD=LIFE PLOTS=(S,H)
   INTERVALS=10 20 30 40 50 53;
   TIME week*arrest(0);
RUN;
```

The resulting hazard estimates and plot in Output 3.15 show only a slight tendency for the hazard to decline in the last interval.

Output 3.15 *Corrected Hazard Estimates and Plot for Recidivism Data*

Interval [Lower, Upper)		Median Standard Error	PDF	PDF Standard Error	Hazard	Hazard Standard Error
0	10	.	0.00324	0.000852	0.003294	0.00088
10	20	.	0.00486	0.00103	0.005153	0.001124
20	30	.	0.00532	0.00108	0.005966	0.001244
30	40	.	0.00532	0.00108	0.006345	0.001322
40	50	.	0.00602	0.00114	0.007692	0.001507
50	53	.	0.00540	0.00202	0.007258	0.002743
53	.	.	.	.	.	.

Evaluated at the Midpoint of the Interval

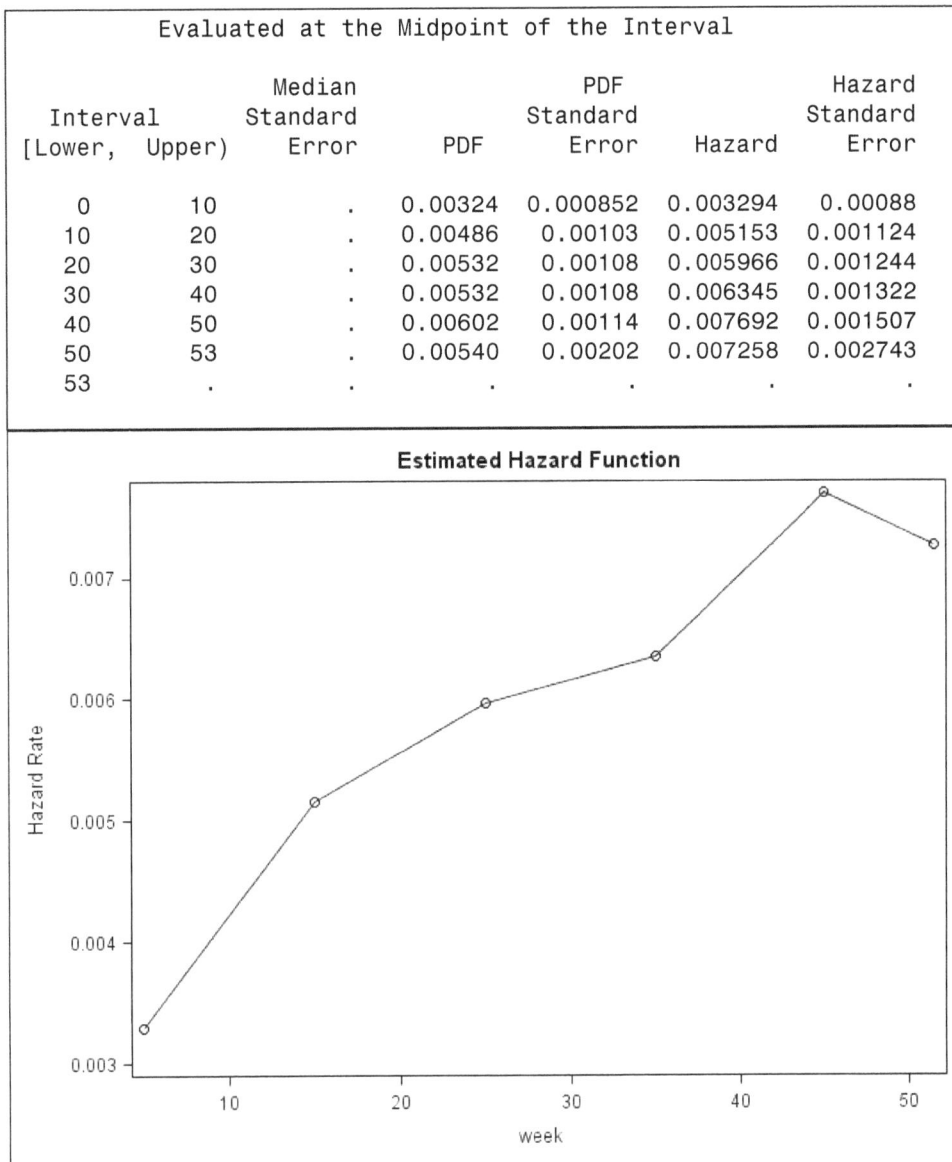

Estimated Hazard Function

LIFE TABLES FROM GROUPED DATA

Although PROC LIFETEST estimates survival probabilities from individual-level data with exact event times, you can also easily construct life tables from published data that provide only

- the boundaries of the intervals
- the number of events in each interval
- the number of censored cases in each interval.

Consider the following survival data for 68 patients from the Stanford
Heart Transplantation Program, as reported by Elandt-Johnson and
Johnson (1980):

Number of Days	Number of Deaths	Number Censored
0-50	16	3
50-100	11	0
100-200	4	2
200-400	5	4
400-700	2	6
700-1000	4	3
1000-1300	1	2
1300-1600	1	3
1600+	0	1

Time is measured from the date of the transplant. Although this sample is
rather small for constructing a life table, it will do fine for illustration. The
trick is to create a separate observation for each of the frequency counts in
the table. For each observation, the value of the time variable (TIME) can
be anywhere within the interval—we'll use the midpoint. A second
variable (STATUS) is created with a value of 1 for the death counts and a
value of 0 for the censored counts. The frequency count (NUMBER) is used
as a weight variable with the FREQ statement in PROC LIFETEST. Here's
the SAS code:

```
DATA;
INPUT time status number;
DATALINES;
25 1 16
25 0 3
75 1 11
75 0 0
150 1 4
150 0 2
300 1 5
300 0 4
550 1 2
550 0 6
850 1 4
850 0 3
1150 1 1
1150 0 2
1450 1 1
```

```
1450 0 3
1700 1 0
1700 0 1
;
PROC LIFETEST METHOD=LIFE INTERVALS=50 100 200 400 700
  1000 1300 1600 PLOTS=(S,H);
   TIME time*status(0);
   FREQ number;
RUN;
```

Two of the data lines (75 0 0, 1700 1 0) are unnecessary because the frequency is 0, but they are included here for completeness. Output 3.16 shows the tabular results. The hazard plot is in Output 3.17.

The most striking fact about this example is the rapid decline in the hazard of death from the origin to about 200 days after surgery. After that, the hazard is fairly stable. This decline is reflected in the Median Residual Lifetime column. At time 0, the median residual lifetime of 257.7 days is an estimate of the median survival time for the entire sample. However, of those patients still alive at 50 days, the median residual lifetime rises to 686.6 days. The median remaining life continues to rise until it reaches a peak of 982.9 days for those who were still alive 400 days after surgery.

Output 3.16   *Life-Table Estimates from Grouped Data*

```
Life Table Survival Estimates

                                          Effective    Conditional
      Interval        Number    Number      Sample      Probability
  [Lower,  Upper)     Failed    Censored     Size        of Failure

      0        50       16         3         66.5          0.2406
     50       100       11         0         49.0          0.2245
    100       200        4         2         37.0          0.1081
    200       400        5         4         30.0          0.1667
    400       700        2         6         20.0          0.1000
    700      1000        4         3         13.5          0.2963
   1000      1300        1         2          7.0          0.1429
   1300      1600        1         3          3.5          0.2857
   1600        .         0         1          0.5          0
```

(*continued*)

Output 3.16 *(continued)*

Interval [Lower,	Upper)	Conditional Probability Standard Error	Survival	Failure	Survival Standard Error	Median Residual Lifetime
0	50	0.0524	1.0000	0	0	257.7
50	100	0.0596	0.7594	0.2406	0.0524	686.6
100	200	0.0510	0.5889	0.4111	0.0608	855.7
200	400	0.0680	0.5253	0.4747	0.0620	910.5
400	700	0.0671	0.4377	0.5623	0.0628	982.9
700	1000	0.1243	0.3939	0.6061	0.0637	779.6
1000	1300	0.1323	0.2772	0.7228	0.0664	.
1300	1600	0.2415	0.2376	0.7624	0.0677	.
1600	.	0	0.1697	0.8303	0.0750	.

Evaluated at the Midpoint of the Interval

Interval [Lower,	Upper)	Median Standard Error	PDF	PDF Standard Error	Hazard	Hazard Standard Error
0	50	140.1	0.00481	0.00105	0.00547	0.001355
50	100	139.4	0.00341	0.000935	0.005057	0.001513
100	200	124.4	0.000637	0.000308	0.001143	0.00057
200	400	363.2	0.000438	0.000186	0.000909	0.000405
400	700	216.3	0.000146	0.0001	0.000351	0.000248
700	1000	236.9	0.000389	0.000175	0.001159	0.000571
1000	1300	.	0.000132	0.000126	0.000513	0.000511
1300	1600	.	0.000226	0.000202	0.001111	0.001096
1600	.	.	.	.	.	.

Output 3.17 *Hazard Plot for Grouped Data*

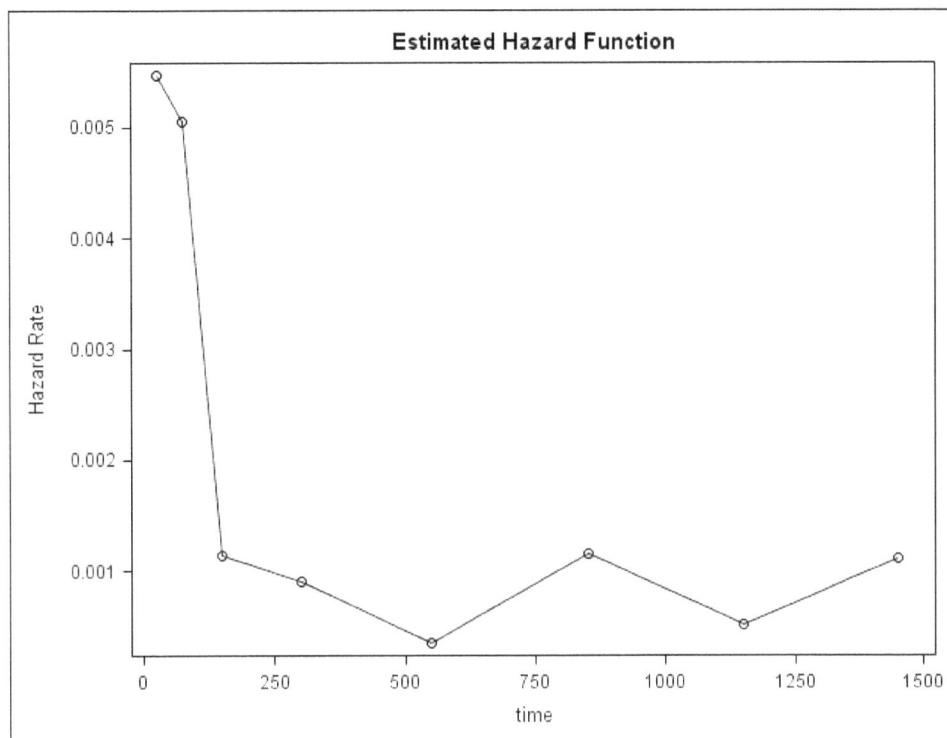

TESTING FOR EFFECTS OF COVARIATES

Besides tests of differences between groups, PROC LIFETEST can test whether quantitative covariates are associated with survival time. Given a list of covariates, PROC LIFETEST produces a test statistic for each one and ignores the others. It also treats them as a set, testing the null hypothesis that they are jointly unrelated to survival time and also testing for certain incremental effects of adding variables to the set. The statistics are generalizations of the log-rank and Wilcoxon tests discussed earlier in this chapter. They can also be interpreted as nonparametric tests of the coefficients of the accelerated failure time model discussed in Chapter 4, "Estimating Parametric Regression Models with PROC LIFEREG."

You can test the same sorts of hypotheses with PROC LIFEREG or PROC PHREG. In fact, the log-rank chi-square reported by PROC LIFETEST is identical to the score statistic given by PROC PHREG for the null hypothesis that all coefficients are 0 (when the data contain no tied event times). However, in most cases, you are better off switching to the regression procedures, for two reasons. First, PROC LIFETEST doesn't give coefficient estimates, so there is no way to quantify the effect of a covariate on survival time. Second, the incremental tests do not really test the effect of each variable controlling for all the others. Instead, you get a test of the effect of each variable controlling for those variables that have already been included. Because you have no control over the order of inclusion, these tests can be misleading. Nevertheless, PROC LIFETEST can be useful for screening a large number of covariates before proceeding to estimate regression models. Because the log-rank and Wilcoxon tests do not require iterative calculations, they use relatively little computer time. (This is also true for the SELECTION=SCORE option in PROC PHREG.)

Let's look at the recidivism data as an example. The covariate tests are invoked by listing the variable names in a TEST statement:

```
PROC LIFETEST DATA=recid;
   TIME week*arrest(0);
   TEST fin age race wexp mar paro prio;
RUN;
```

Output 3.18 shows selections from the output. I have omitted the Wilcoxon statistics because they are nearly identical to the log-rank statistics for this example. I also omitted the variance-covariance matrix for the statistics because it is primarily useful as input to other analyses.

Output 3.18 *Covariate Tests for the Recidivism Data*

```
        Univariate Chi-Squares for the LOG RANK Test

                   Test                              Pr >
    Variable    Statistic    Variance   Chi-Square   Chi-Square

    FIN          10.4256     28.4744      3.8172      0.0507
    AGE            233.2      4305.3     12.6318      0.0004
    RACE         -2.7093     12.8100      0.5730      0.4491
    WEXP         16.4141     27.3305      9.8580      0.0017
    MAR           7.1773     13.1535      3.9164      0.0478
    PARO          2.9471     26.7927      0.3242      0.5691
    PRIO          -108.8       812.3     14.5602      0.0001


    Forward Stepwise Sequence of Chi-Squares for the LOG RANK Test

                               Pr >       Chi-Square      Pr >
    Variable    DF Chi-Square  Chi-Square  Increment   Increment

    PRIO         1  14.5602    0.0001      14.5602      0.0001
    AGE          2  25.4905    0.0001      10.9303      0.0009
    FIN          3  28.8871    0.0001       3.3966      0.0653
    MAR          4  31.0920    0.0001       2.2050      0.1376
    RACE         5  32.4214    0.0001       1.3294      0.2489
    WEXP         6  33.2800    0.0001       0.8585      0.3541
    PARO         7  33.3828    0.0001       0.1029      0.7484
```

The top panel shows that age at release (AGE), work experience (WEXP), and number of prior convictions (PRIO) have highly significant associations with time to arrest. The effects of marital status (MAR) and financial aid (FIN) are more marginal, while race and parole status (PARO) are apparently unrelated to survival time. The signs of the log-rank test statistics tell you the direction of the relationship. The negative sign for PRIO indicates that inmates with more prior convictions tend to have shorter times to arrest. On the other hand, the positive coefficient for AGE indicates that older inmates have longer times to arrest. As already noted, none of these tests controls or adjusts for any of the other covariates.

The lower panel displays results from a forward inclusion procedure. PROC LIFETEST first finds the variable with the highest chi-square statistic in the top panel—in this case, PRIO—and puts it in the set to be tested. Because PRIO is the only variable in the set, the results for PRIO are the same in both panels. Then PROC LIFETEST finds the variable that produces the largest increment in the joint chi-square for the set of two variables—in this case, AGE. The joint chi-square of 25.49 in line 2 tests the null hypothesis that the coefficients of AGE and PRIO in an accelerated-failure time model are both 0. The chi-square increment of 10.93 is merely the difference between the joint chi-square in lines 1 and 2. It is a test of the null hypothesis that the coefficient for AGE is 0 when PRIO is controlled. On the other hand, there is no test for the effect of PRIO controlling for AGE.

This process is repeated until all the variables are added. For each variable, we get a test of the hypothesis that the variable is unrelated to survival time controlling for all the variables above it (but none of the variables below it). For variables near the end of the sequence, the incremental chi-square values are likely to be similar to what you might find with PROC LIFEREG or PROC PHREG. For variables near the beginning of the sequence, however, the results can be quite different.

For this example, the forward inclusion procedure leads to some substantially different conclusions from the univariate procedure. While WEXP has a highly significant effect on survival time when considered by itself, there is no evidence of such an effect when other variables are controlled. The reason is that work experience is moderately correlated with age and the number of prior convictions, both of which have substantial effects on survival time. Marital status also loses its statistical significance in the forward inclusion test.

What is the relationship between the STRATA statement and the TEST statement? For a dichotomous variable like FIN, the statement TEST FIN is a possible alternative to STRATA FIN. Both produce a test of the null hypothesis that the survivor functions are the same for the two categories of FIN. In fact, if there are no ties in the data (no cases with exactly the same event time), the two statements will produce identical chi-square statistics and p-values. In the presence of ties, however, STRATA and TEST use somewhat different formulas, which may result in slight differences in the p-values. (If you're interested in the details, see Collett, 2003.) In the recidivism data, for example, the 114 arrests occurred at only 49 unique arrest times, so the number of ties was substantial. The STRATA statement produces a log-rank chi-square of 3.8376 for a p-value of .0501 and a Wilcoxon chi-square of 3.7495 for a p-value of .0528. The TEST statement produces a log-rank chi-square of 3.8172 for a p-value of .0507 and a Wilcoxon chi-square of 3.7485 for a p-value of .0529. Obviously the differences are minuscule in this case.

Other considerations should govern the choice between the STRATA and TEST statements. While the STRATA statement produces separate tables and graphs of the survivor function for the two groups, the TEST statement produces only the single table and graph for the entire sample. With the TEST statement, you can test for the effects of many dichotomous variables with a single statement, but the STRATA statement requires a new PROC LIFETEST step for each variable tested. Of course, if a variable has more than two values, the STRATA statement treats each value as a separate group while the TEST statement treats the variable as a quantitative measure.

What happens when you include both a STRATA statement and a TEST statement? Adding a TEST statement has no effect whatever on the results from the STRATA statement. This fact implies that the hypothesis test produced by the STRATA statement in no way controls for the variables listed in the TEST statement. On the other hand, the TEST statement can produce quite different results, depending on whether you also have a STRATA statement. When you have a STRATA statement, the log-rank and Wilcoxon statistics produced by the TEST statement are first calculated within strata and then averaged across strata. In other words, they are stratified statistics that control for whatever variable or variables are listed in the STRATA statement. Suppose, for example, that for the myelomatosis data, we want to test the effect of the treatment while controlling for renal functioning. We can submit these statements:

```
PROC LIFETEST DATA=myel;
   TIME dur*status(0);
   STRATA renal;
   TEST treat;
RUN;
```

The resulting log-rank chi-square for TREAT is 5.791 with a *p*-value of .016. This result is in sharp contrast with the unstratified chi-square of only 1.3126 that we saw earlier in this chapter (Output 3.7).

An alternative way to get a stratified test comparing survivor functions for different groups is to use the GROUP option in the STRATA statement, as follows:

```
PROC LIFETEST DATA=myel;
   TIME dur*status(0);
   STRATA renal / GROUP=treat;
RUN;
```

This produces the same log-rank chi-square for TREAT of 5.791 with a *p*-value of .016. But the advantage of this approach is that the GROUP variable (unlike a variable in the TEST statement) can have more than two categories. As we'll see in Chapter 5, "Estimating Cox Regression Models with PROC PHREG," you can also get nearly identical results using PROC PHREG with stratification and the score test.

LOG SURVIVAL AND SMOOTHED HAZARD PLOTS

PROC LIFETEST produces two other plots that give useful information about the shape of the hazard function, the log-survival (LS) plot and the log-log survival (LLS) plot. In this section, we'll see how these plots help determine whether the hazard function can be accurately described by certain parametric models discussed in Chapter 2. We'll also see how to get smoothed estimates of the hazard function with ungrouped data.

Suppose we specify PLOTS=(S, LS, LLS) in the PROC LIFETEST statement. The S gives us the now-familiar survival curve. LS produces a plot of $-\log \hat{S}(t)$ versus t. To explain how this plot is useful, we need a little background. From equation (2.6), one can readily see that

$$-\log S(t) = \int_0^t h(u)du$$

Because of this relationship, the log survivor function is commonly referred to as the *cumulative hazard function*, frequently denoted by $\Lambda(t)$. Now, if $h(t)$ is a constant with a value of λ (which implies an exponential distribution for event times), then the cumulative hazard function is just $\Lambda(t) = \lambda t$. This result implies that a plot of $-\log \hat{S}(t)$ versus t should yield a straight line with an origin at 0. Moreover, an examination of the log-survival plot can tell us whether the hazard is constant, increasing, or decreasing with time.

Output 3.19 displays the LS plot for the myelomatosis data. Instead of a straight line, the graph appears to increase at a *decreasing* rate. This fact suggests that the hazard is *not* constant but rather declines with time. If the plot had curved upward rather than downward, it would suggest that the hazard was *increasing* with time. Of course, because the sample size is quite small, caution is advisable in drawing any conclusions. A formal test, such as the one described in the next chapter, might not show a significant decrease in the hazard.

Output 3.19 *Log-Survival Plot for Myelomatosis Data*

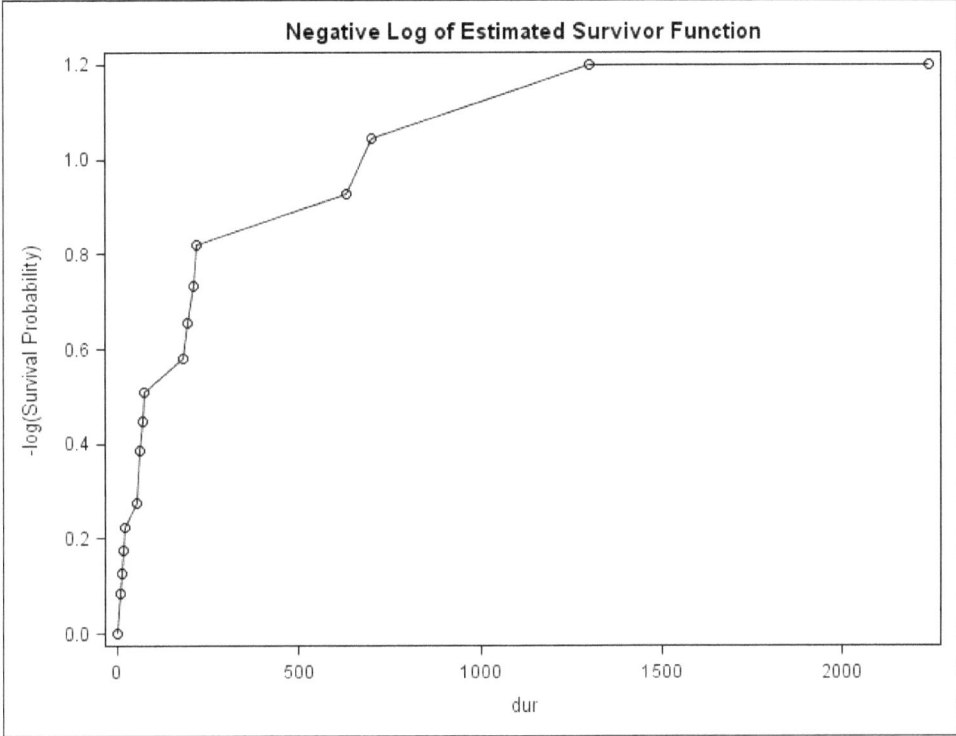

The LLS keyword produces a plot of $\log[-\log \hat{S}(t)]$ versus log t. If survival times follow a Weibull distribution, which has a hazard given by $\log h(t) = \alpha + \beta \log t$, then the log-log survival plot (log cumulative hazard plot) should be a straight line with a slope of β. Examining Output 3.20, we see a rather rough plot with a slight tendency to turn downward at later times. Again, however, the data are so sparse that this is probably not sufficient evidence for rejecting the Weibull distribution. In Chapter 4, we'll see how to construct similar plots for other distributions such as the log-normal and log-logistic.

Output 3.20 *Log-Log Survival Plot for Myelomatosis Data*

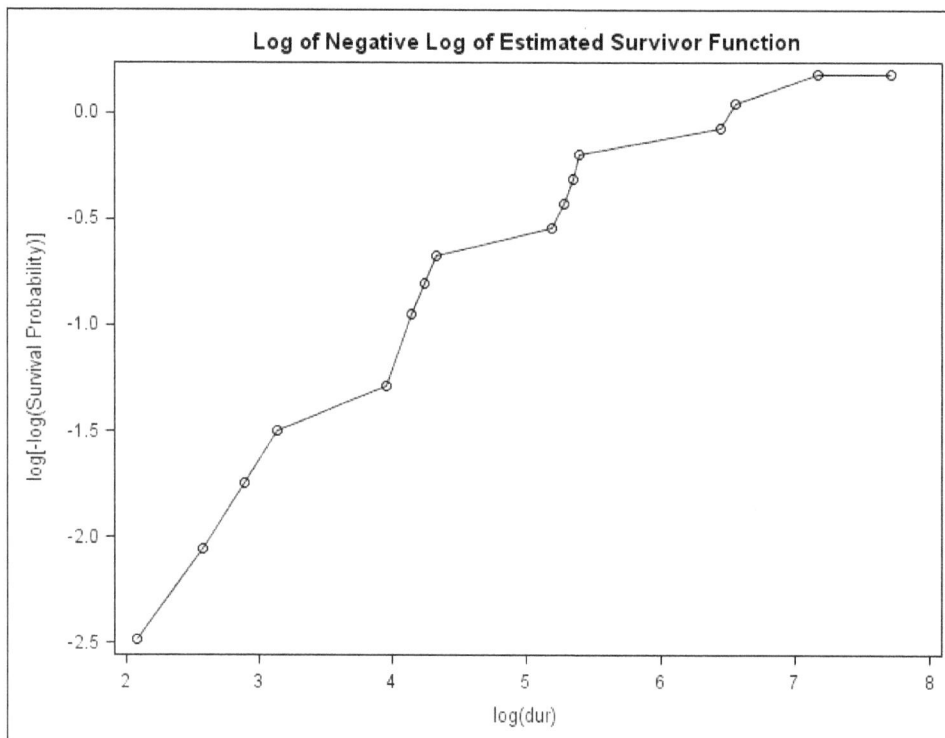

Although graphs based on transformations of the survivor function are certainly useful, they are ultimately rather frustrating. What we really want to see is a graph of the hazard function. We got that from the life-table method, but at the cost of grouping the event times into arbitrarily chosen intervals. While it's possible to estimate hazards from ungrouped data, the estimates usually vary so wildly from one time to the next as to be almost useless. There are several ways to smooth these estimates by calculating some sort of moving average. One method, known as *kernel smoothing*, has been shown to have good properties for hazard functions (Ramlau-Hansen, 1983). Using this method, PROC LIFETEST (SAS 9.2 and later) can produce smoothed graphs of hazard functions with the ODS version of the PLOT option. Here's how to do it for the recidivism data:

```
ODS GRAPHICS ON;
PROC LIFETEST DATA=recid PLOTS=H;
  TIME week*arrest(0);
RUN;
ODS GRAPHICS OFF;
```

Output 3.21 shows the smoothed hazard function. The graph bears some resemblance to the grouped hazard plot in Output 3.14, although here we see a rather pronounced plateau between about 20 and 30 weeks and no downturn at the end.

Output 3.21   *Smoothed Hazard Function Estimate for Recidivism Data*

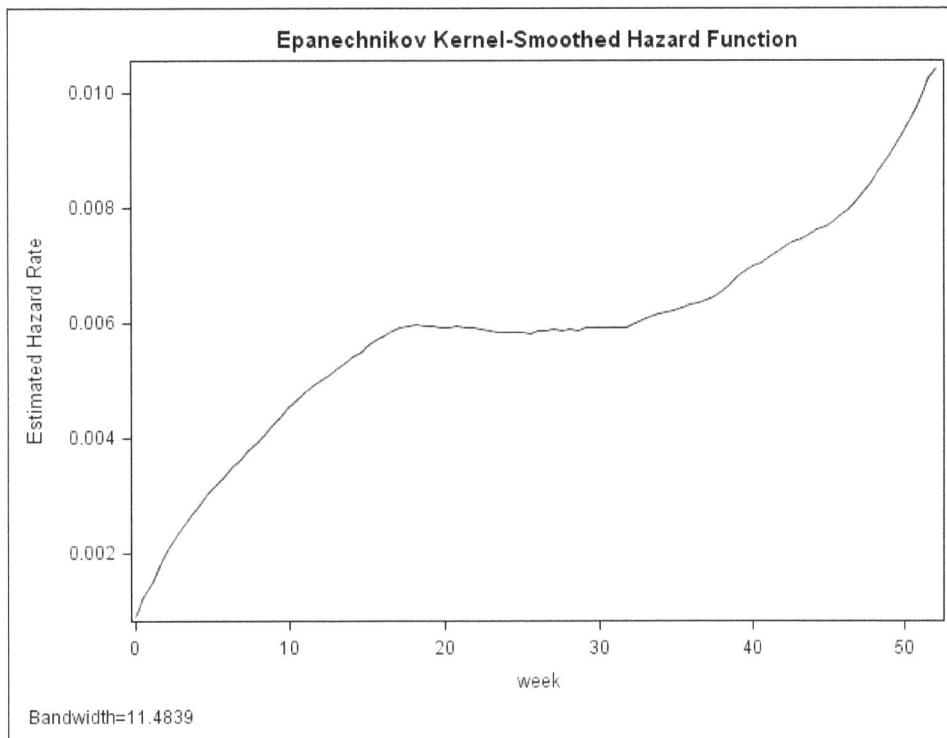

Notice the Bandwidth=11.4839 in the lower left corner. This means that, when calculating the hazard for any specified point in time, the smoothing function only uses data within 11.5 weeks on either side of the time point. The bandwidth has a big impact on the appearance of the graph: the larger the bandwidth, the smoother the graph. By default, PROC LIFETEST selects the bandwidth using a method that has certain optimality properties. But sometimes you may want to experiment with different bandwidths. Here's how to force the bandwidth to be 5 weeks:

```
PROC LIFETEST DATA=recid PLOTS=H(BW=5);
```

As Output 3.22 shows, the resulting hazard function is much choppier.

Output 3.22 *Smoothed Hazard Function Estimate with Smaller Bandwidth*

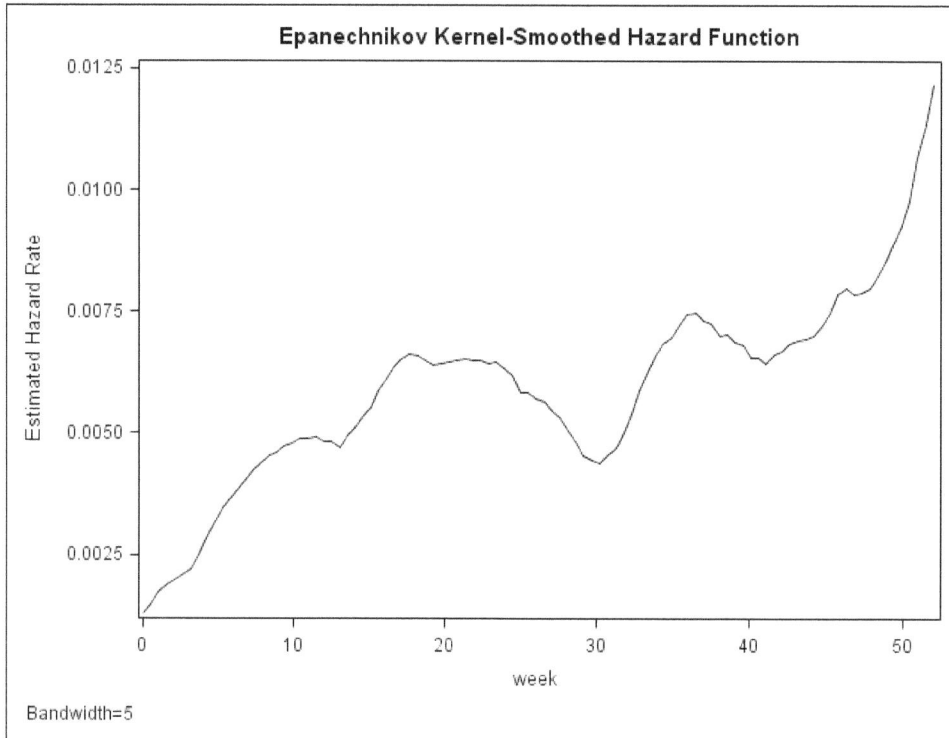

If you want confidence limits around the hazard function, you can get them using the CL option:

```
PROC LIFETEST DATA=recid PLOTS=H(CL);
```

However, as we see in Output 3.23, the hazard curve is greatly compressed because the confidence limits are so wide at the earliest and latest times.

Output 3.23 *Smoothed Hazard Function with Confidence Limits*

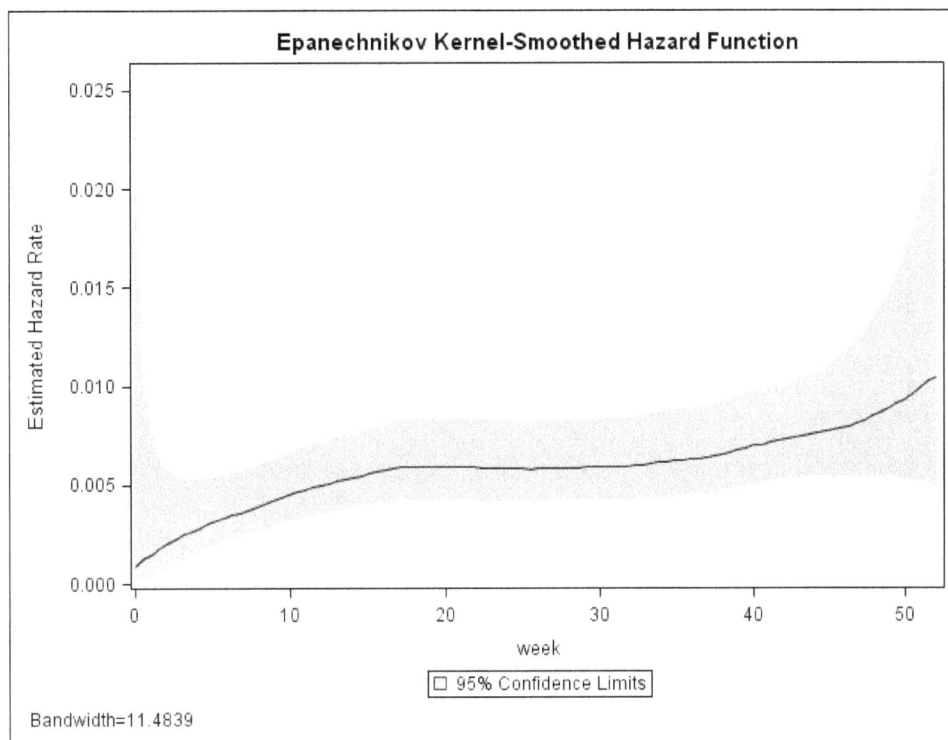

Epanechnikov Kernel-Smoothed Hazard Function

(Bandwidth=11.4839)

CONCLUSION

PROC LIFETEST is a useful procedure for preliminary analysis of survival data and for testing simple hypotheses about differences in survival across groups. For experimental studies, PROC LIFETEST (with the STRATA statement) gives tests that are analogous to a one-way analysis of variance. But the procedure is not adequate for two-factor designs because there is no way to test for interactions. Similarly, while the TEST statement in PROC LIFETEST may be useful for screening large numbers of quantitative covariates, it is not adequate for examining the effects of variables controlling for other covariates. In most cases, therefore, you will need to move to the estimation of regression models with PROC LIFEREG or PROC PHREG.

It is also important to recognize that survival curves and their associated hazard functions can be misleading when the sample is heterogeneous. As explained in Chapter 8, uncontrolled heterogeneity tends to make hazard functions look as though they are declining, even when there is no real decline for any individual in the sample. As we will see in Chapter 5, one way to reduce the effect of heterogeneity is to estimate and plot baseline survivor functions after fitting Cox regression models.

CHAPTER **4**
Estimating Parametric Regression Models with PROC LIFEREG

INTRODUCTION

The LIFEREG procedure produces estimates of parametric regression models with censored survival data using the method of maximum likelihood. To some degree, PROC LIFEREG has been eclipsed by the PHREG procedure, which does *semiparametric* regression analysis using a method known as *partial* likelihood. The reasons for PROC PHREG's popularity will become apparent in the next chapter. PROC LIFEREG is by no means obsolete, however. It can do some things better than PROC PHREG, and it can do other things that PROC PHREG can't do at all:

- PROC LIFEREG accommodates left censoring and interval censoring. PROC PHREG allows only right censoring.
- With PROC LIFEREG, you can test certain hypotheses about the shape of the hazard function. PROC PHREG gives you only nonparametric estimates of the *survivor* function, which can be difficult to interpret.

- If the shape of the survival distribution is known, PROC LIFEREG produces more efficient estimates (with smaller standard errors) than PROC PHREG.

- PROC LIFEREG can easily generate predicted event times for any specified set of covariate values. This is more difficult with PHREG, and often impossible.

PROC LIFEREG's greatest limitation is that it does not handle time-dependent covariates, something at which PROC PHREG excels.

THE ACCELERATED FAILURE TIME MODEL

The class of regression models estimated by PROC LIFEREG is known as the *accelerated failure time* (AFT) model. In its most general form, the AFT model describes a relationship between the survivor functions of any two individuals. If $S_i(t)$ is the survivor function for individual i, then for any other individual j, the AFT model holds that

$$S_j(t) = S_i(\phi_{ij}t) \text{ for all t} \tag{4.1}$$

where ϕ_{ij} is a constant that is specific to the pair (i, j). This model says, in effect, that what makes one individual different from another is the rate at which they age. A good example is the conventional wisdom that a year for a dog is equivalent to 7 years for a human. This relationship can be represented by equation (4.1), with S_i being the survival probability for a dog, S_j the survival probability for a human, and $\phi_{ij} = 7$.

What PROC LIFEREG actually estimates is a special case of this model that is quite similar in form to an ordinary linear regression model. Let T_i be a random variable denoting the event time for the ith individual in the sample, and let $x_{i1}, ..., x_{ik}$ be the values of k covariates for that same individual. The model is then

$$\log T_i = \beta_0 + \beta_1 x_{i1} + ... + \beta_k x_{ik} + \sigma\varepsilon_i \tag{4.2}$$

where ε_i is a random disturbance term, and $\beta_0, ..., \beta_k$, and σ are parameters to be estimated. Exponentiating both sides of equation (4.2) gives an alternative way of expressing the model:

$$T_i = \exp(\beta_0 + \beta_1 x_{i1} + ... + \beta_k x_{ik} + \sigma\varepsilon_i)$$

So far, the only differences between the model in equation (4.2) and the usual linear regression model are that there is a σ before the ε and the dependent variable is logged. The σ can be omitted, which requires that the variance of ε be allowed to vary from one data set to another. But it is simpler to fix the variance of ε at some standard value (for example, 1.0) and let σ change in value to accommodate changes in the disturbance variance. (This notational strategy could also be used for linear models.) As for the log transformation of T, its main purpose is to ensure that predicted values of T are positive, regardless of the values of the x's and the β's.

In a linear regression model, it is typical to assume that ε_i has a normal distribution with a mean and variance that is constant over i, and that the ε's are independent across observations. One member of the AFT class, the *log-normal model*, has exactly these assumptions. It is called the log-normal model because if log T has a normal distribution, then T has a log-normal distribution. Other AFT models allow distributions for ε besides the normal distribution but retain the assumptions of constant mean and variance, as well as independence across observations. We will consider these alternative models in some detail, but for the moment let's stick with the log-normal.

If there are no censored data, we can readily estimate this model by ordinary least squares (OLS). Simply create a new variable in a DATA step that is equal to the log of the event time and use the REG procedure with the transformed variable as the dependent variable. This process yields best linear unbiased estimates of the β coefficients, regardless of the shape of the distribution of ε. If ε is normal, the OLS estimates will also be maximum likelihood estimates and will have minimum variance among all estimators, both linear and nonlinear. In fact, the coefficients and standard errors produced by PROC REG will be identical to those produced by PROC LIFEREG with a log-normal specification.

But survival data typically have at least some censored observations, and these are difficult to handle with OLS. Instead, we can use maximum likelihood estimation. Later, this chapter examines the mathematics of maximum likelihood (ML) for censored regression models in some detail. First, let's look at an example of how a regression application is set up with PROC LIFEREG and what results it produces. In the section **The Life Table Method** in Chapter 3, "Estimating and Comparing Survival Curves with PROC LIFETEST," I described the recidivism data set in which 432 inmates were followed for 1 year after release. That data set is used throughout this chapter, so you may want to reread the earlier description (or see Appendix 2, "Data Sets").

In this recidivism example, the variable WEEK contains the week of the first arrest or censoring. The variable ARREST is equal to 1 if WEEK is uncensored or 0 if right censored. There are seven covariates. To estimate the log-normal model, we specify

```
PROC LIFEREG DATA=recid;
  MODEL week*arrest(0)=fin age race wexp mar paro prio
       / DISTRIBUTION=LNORMAL;
RUN;
```

Note that WEEK*ARREST(0) in the MODEL statement follows the same syntax as the TIME statement in PROC LIFETEST. However, it is now followed by an equal sign and a list of covariates. The slash (/) separates the variable list from the specification of options, of which there are several possibilities. Here, we have merely indicated our choice of the log-normal distribution.

Output 4.1 displays the results. The output first provides some preliminary information: the names of the time and censoring variables, the values that correspond to censoring, and the numbers of cases with each type of censoring. Then the output shows that the log-likelihood for the model is −322.6946. This is an important statistic that we will use later to test various hypotheses.

The table labeled "Fit Statistics" gives four modifications of the log-likelihood that may be used in assessing model fit. The first is simply the log-likelihood multiplied by −2. The second, Akaike's information criterion (AIC), is a modification of the −2 log-likelihood that penalizes models for having more covariates. Specifically,

$$AIC = -2\log L + 2k$$

where k is the number of covariates. The next statistic (AICC) is a "corrected" version of the AIC that may have better behavior in small samples:

$$AICC = AIC + \frac{2k(k+1)}{n-k-1}$$

Finally, the Bayesian information criterion (also known as Schwarz's criterion) gives a more severe penalization for additional covariates, at least for most applications:

$$BIC = -2\log L + k\log n$$

All three of these penalized statistics can be used to compare models with different sets of covariates. The models being compared do not have to be nested, in the sense of one model being a special case of another. However, these statistics cannot be used to construct a formal hypothesis test, so the comparison is only informal. For all three statistics, smaller values mean a better fit. But keep in mind that the overall magnitude of these statistics depends heavily on sample size.

For this example, the "Type III Analysis of Effects" table is completely redundant with the table that follows. But the Type III table will be important later when we consider CLASS variables.

Finally, we get a table of estimated coefficients, their standard errors, chi-square statistics for the null hypothesis that each coefficient is 0, and *p*-values associated with those statistics. The chi-squares are calculated by dividing each coefficient by its estimated standard error and squaring the result.

Output 4.1　*Results from Fitting a Log-Normal Model to the Recidivism Data*

```
                    The LIFEREG Procedure

                     Model Information

        Data Set                      RECID
        Dependent Variable            Log(week)
        Censoring Variable            arrest
        Censoring Value(s)                0
        Number of Observations          432
        Noncensored Values              114
        Right Censored Values           318
        Left Censored Values              0
        Interval Censored Values          0
        Name of Distribution        Lognormal
        Log Likelihood            -322.6945851


          Number of Observations Read       432
          Number of Observations Used       432

                    Fit Statistics

        -2 Log Likelihood               645.389
        AIC (smaller is better)         663.389
        AICC (smaller is better)        663.816
        BIC (smaller is better)         700.005

Algorithm converged.
```

(continued)

Output 4.1 *(continued)*

```
                  Type III Analysis of Effects

                                    Wald
            Effect       DF    Chi-Square    Pr > ChiSq

            fin           1      4.3657        0.0367
            age           1      2.9806        0.0843
            race          1      1.8824        0.1701
            wexp          1      2.2466        0.1339
            mar           1      2.4328        0.1188
            paro          1      0.1092        0.7411
            prio          1      5.8489        0.0156


         Analysis of Maximum Likelihood Parameter Estimates
```

Parameter	DF	Estimate	Standard Error	95% Confidence Limits		Chi-Square	Pr > ChiSq
Intercept	1	4.2677	0.4617	3.3628	5.1726	85.44	<.0001
fin	1	0.3428	0.1641	0.0212	0.6645	4.37	0.0367
age	1	0.0272	0.0158	-0.0037	0.0581	2.98	0.0843
race	1	-0.3632	0.2647	-0.8819	0.1556	1.88	0.1701
wexp	1	0.2681	0.1789	-0.0825	0.6187	2.25	0.1339
mar	1	0.4604	0.2951	-0.1181	1.0388	2.43	0.1188
paro	1	0.0559	0.1691	-0.2756	0.3873	0.11	0.7411
prio	1	-0.0655	0.0271	-0.1186	-0.0124	5.85	0.0156
Scale	1	1.2946	0.0990	1.1145	1.5038		

Two variables meet the .05 criterion for statistical significance: FIN (whether the inmate received financial aid) and PRIO (number of prior convictions). The signs of the coefficients tell us the direction of the relationship. The positive coefficient for FIN indicates that those who received financial aid had longer times to arrest than those who did not. The negative coefficient for PRIO indicates that additional convictions were associated with shorter times to arrest. As in any regression procedure, these coefficients adjust or control for the other covariates in the model.

The numerical magnitudes of the coefficients are not very informative in the reported metrics, but a simple transformation leads to a more intuitive interpretation. For a 1-0 variable like FIN, if we simply take e^β, we get the estimated ratio of the expected (mean) survival times for the two groups. Thus, $e^{.3428} = 1.41$. Therefore, controlling for the other covariates, the expected time to arrest for those who received financial aid is 41 percent greater than for those who did not receive financial aid. (This statement also applies to the median time to arrest, or any other percentile for that matter.) For a quantitative variable like PRIO, we can use the transformation $100(e^\beta - 1)$, which gives the percent change in the expected survival time for each one-unit increase in the variable. Thus, $100(\exp(-.0655) - 1) = -6.34$. According to the model, then, each additional prior conviction is associated with a 6.34 percent decrease in expected time to arrest, holding other covariates constant. We can also interpret the coefficients for any of the other AFT models discussed in this chapter in this way.

The output line labeled SCALE is an estimate of the σ parameter in equation (4.2), along with its estimated standard error. For some distributions, changes in the value of this parameter can produce qualitative differences in the shape of the hazard function. For the log-normal model, however, changes in σ merely compress or stretch the hazard function.

ALTERNATIVE DISTRIBUTIONS

In ordinary linear regression, the assumption of a normal distribution for the disturbance term is routinely invoked for a wide range of applications. Yet PROC LIFEREG allows for four additional distributions for ε: extreme value (2 parameter), extreme value (1 parameter), log-gamma, and logistic. For each of these distributions, there is a corresponding distribution for T:

Distribution of ε	Distribution of T
extreme value (2 par.)	Weibull
extreme value (1 par.)	exponential
log-gamma	gamma
logistic	log-logistic
normal	log-normal

Incidentally, all AFT models are named for the distribution of T rather than for the distribution of ε or log T. Because the logistic and normal lead to the log-logistic and log-normal, you might expect that the gamma will lead to the log-gamma. But it is just the reverse. This is one of those unfortunate inconsistencies in terminology that we just have to live with.

What is it about survival analysis that makes these alternatives worth considering? The main reason for allowing other distributions is that they have different implications for *hazard functions* that may, in turn, lead to different substantive interpretations. The remainder of this section explores each of these alternatives in some detail.

The Exponential Model

The simplest model that PROC LIFEREG estimates is the exponential model, invoked by DISTRIBUTION=EXPONENTIAL in the MODEL statement. This model specifies that ε has a standard extreme-value distribution and constrains $\sigma = 1$. If ε has an extreme-value distribution, then log T also has an extreme-value distribution, conditional on the covariates. This implies that T itself has an exponential distribution, which is why we call it the *exponential model*. The standard extreme value distribution is also known as a *Gumbel distribution* or a *double exponential distribution*. It has a p.d.f. of $f(\varepsilon) = \exp[\varepsilon - \exp(\varepsilon)]$. Like the normal distribution, this is a unimodal distribution defined on the entire real line. Unlike the normal, however, it is not symmetrical, being slightly skewed to the left.

As we saw in Chapter 2, "Basic Concepts of Survival Analysis," an exponential distribution for T corresponds to a *constant hazard function*, which is the most characteristic feature of this model. However, equation (2.12) expresses the exponential regression model as

$$\log h(t) = \beta_0^\bullet + \beta_1^\bullet x_1 + \ldots + \beta_k^\bullet x_k \tag{4.3}$$

where the $\bullet$'s have been added to distinguish these coefficients from those in equation (4.2). Although the dependent variable in equation (4.2) is the log of time, in equation (4.3) it is the log of the hazard. It turns out that the two models are completely equivalent. Furthermore, there is a simple relationship between the coefficients in equation (4.2) and equation (4.3): namely, that $\beta_j = -\beta_j^\bullet$ for all j.

The change in signs makes intuitive sense. If the hazard is high, then events occur quickly and survival times are short. On the other hand, when the hazard is low, events are unlikely to occur and survival times are long. It is important to be able to shift back and forth between these two ways of expressing the model so that you can compare results across different computer programs. In particular, because PROC PHREG reports coefficients in log-hazard form, we need to make the conversion in order to compare PROC LIFEREG output with PROC PHREG output.

You may wonder why there is no disturbance term in equation (4.3) (a characteristic it shares with the more familiar logistic regression model). No disturbance term is needed because there is implicit random variation in the relationship between $h(t)$, the unobserved hazard, and the observed event time T. Even if two individuals have exactly the same covariate values (and, therefore, the same hazard), they will not have the same event time. Nevertheless, in Chapter 8, "Heterogeneity, Repeated Events, and Other Topics," we will see that there have been some attempts to add a disturbance term to models like this to represent *unobserved heterogeneity*.

Output 4.2 shows the results of fitting the exponential model to the recidivism data. Comparing this with the log-normal results in Output 4.1, we see some noteworthy differences. The coefficient for AGE is about twice as large in the exponential model, and its *p*-value declines from .08 to .01. Similarly, the coefficient for PRIO increases somewhat in magnitude, and its *p*-value also goes down substantially. On the other hand, the *p*-value for FIN increases to slightly above the .05 level.

Output 4.2   *Exponential Model Applied to Recidivism Data*

```
         Analysis of Maximum Likelihood Parameter Estimates

                        Standard   95% Confidence     Chi-
Parameter       DF Estimate   Error       Limits     Square Pr > ChiSq

Intercept        1   4.0507  0.5860   2.9021   5.1993  47.78    <.0001
fin              1   0.3663  0.1911  -0.0083   0.7408   3.67   0.0553
age              1   0.0556  0.0218   0.0128   0.0984   6.48   0.0109
race             1  -0.3049  0.3079  -0.9085   0.2986   0.98   0.3220
wexp             1   0.1467  0.2117  -0.2682   0.5617   0.48   0.4882
mar              1   0.4270  0.3814  -0.3205   1.1745   1.25   0.2629
paro             1   0.0826  0.1956  -0.3007   0.4660   0.18   0.6726
prio             1  -0.0857  0.0283  -0.1412  -0.0302   9.15   0.0025
Scale            0   1.0000  0.0000   1.0000   1.0000
Weibull Shape    0   1.0000  0.0000   1.0000   1.0000


              Lagrange Multiplier Statistics

        Parameter      Chi-Square     Pr > ChiSq
        Scale            24.9302         <.0001
```

Clearly, the choice of model can make a substantive difference. Later, this chapter considers some criteria for choosing among these and other models. Notice that the SCALE parameter σ is forced equal to 1.0 which, as noted above, is what distinguishes the exponential model from the Weibull model. The line labeled "Weibull Shape" will be explained when we discuss the Weibull model.

The "Lagrange Multiplier Statistics" table reports a chi-square of 24.9302 with a *p*-value less than .0001. This is a 1 degree-of-freedom test for the null hypothesis that $\sigma = 1$. Here the null hypothesis is soundly rejected, indicating that the hazard function is *not* constant over time. While this might suggest that the log-normal model is superior, things are not quite that simple. There are other models to consider as well.

The Weibull Model

The Weibull model is a slight modification of the exponential model, with big consequences. By specifying DISTRIBUTION=WEIBULL in the MODEL statement, we retain the assumption that ε has a standard extreme-value distribution, but we relax the assumption that $\sigma = 1$. When $\sigma > 1$, the hazard decreases with time. When $.5 < \sigma < 1$, the hazard is increasing at a decreasing rate. When $0 < \sigma < .5$, the hazard is increasing at an increasing rate. And when $\sigma = .5$, the hazard function is an increasing straight line with an origin at 0. Graphs of these hazard functions appear in Figure 2.3 (with the α in the figure equal to $1/\sigma - 1$).

We call this the Weibull model because T has a Weibull distribution, conditional on the covariates. The Weibull distribution has long been the most popular parametric model in the biostatistical literature, for two reasons. First, it has a relatively simple survivor function that is easy to manipulate mathematically:

$$S_i(t) = \exp\left\{ -\left[t_i e^{-\beta \mathbf{x}_i} \right]^{\frac{1}{\sigma}} \right\}$$

where $\mathbf{x}_i$ is a vector of the covariate values and $\boldsymbol{\beta}$ is a vector of coefficients. Second, in addition to being an AFT model, the Weibull model is also a proportional hazards model. This means that its coefficients (when suitably transformed) can be interpreted as relative hazard ratios. In fact, the Weibull model (and its special case, the exponential model) is the only model that is simultaneously a member of both these classes.

As with the exponential model, there is an exact equivalence between the log-hazard form of the model

$$\log h(t) = \alpha \log t + \beta_0^* + \beta_1^* x_1 + \ldots + \beta_k^* x_k$$

and the log-survival time model

$$\log T_i = \beta_0 + \beta_1 x_1 + \ldots + \beta_k x_k + \sigma \varepsilon$$

The relationship between the parameters is slightly more complicated, however. Specifically, for the Weibull model

$$\beta_j^* = \frac{-\beta_j}{\sigma} \text{ for } j = 1, \ldots, k$$

and $\alpha = (1/\sigma) - 1$. Because $\beta_j = 0$ if and only if $\beta_j^{\bullet} = 0$, a test of the null hypothesis that a coefficient is 0 will be the same regardless of which form you use. On the other hand, standard errors and confidence intervals for coefficients in the log-survival time format are not so easily converted to the log-hazard format. Collett (2003) gives formulas for accomplishing this.

Output 4.3 shows the results from fitting the Weibull model to the recidivism data. Compared with the exponential model in Output 4.2, the coefficients are all somewhat attenuated. But the standard errors are also smaller, so the chi-square statistics and p-values are hardly affected at all. Furthermore, if we convert the coefficients to the log-hazard format by changing the sign and dividing by $\hat{\sigma}$ (the Scale estimate of .7124 in the output), we get

FIN	-0.382
AGE	-0.057
RACE	0.316
WEXP	-0.150
MAR	-0.437
PARO	-0.083
PRIO	0.092.

These coefficients are much closer to the log-hazard coefficients for the exponential model (which differ only in sign from the log-survival time coefficients).

Output 4.3 *Weibull Model Applied to Recidivism Data*

Parameter	DF	Estimate	Standard Error	95% Confidence Limits		Chi-Square	Pr > ChiSq
Intercept	1	3.9901	0.4191	3.1687	4.8115	90.65	<.0001
fin	1	0.2722	0.1380	0.0018	0.5426	3.89	0.0485
age	1	0.0407	0.0160	0.0093	0.0721	6.47	0.0110
race	1	-0.2248	0.2202	-0.6563	0.2067	1.04	0.3072
wexp	1	0.1066	0.1515	-0.1905	0.4036	0.49	0.4820
mar	1	0.3113	0.2733	-0.2244	0.8469	1.30	0.2547
paro	1	0.0588	0.1396	-0.2149	0.3325	0.18	0.6735
prio	1	-0.0658	0.0209	-0.1069	-0.0248	9.88	0.0017
Scale	1	0.7124	0.0634	0.5983	0.8482		
Weibull Shape	1	1.4037	0.1250	1.1789	1.6713		

Because $\hat{\sigma}$ (labeled "Scale" in Output 4.3) is between 0 and 1, we conclude that the hazard is increasing at a decreasing rate. We can also calculate $\hat{\alpha} = (1/.7124) - 1 = 0.4037$, which is the coefficient for log t in the log-hazard model. Because both the dependent and independent variables are logged, this coefficient can be interpreted as follows: a 1 percent increase in time since release is associated with a 0.40 percent increase in the hazard for arrest.

The last line is labeled "Weibull Shape." This is not an independent parameter but merely the reciprocal of the Shape parameter (that is, $1/\hat{\sigma}$). It is included in the output because some statisticians prefer this way of parameterizing the model.

The Log-Normal Model

Although we have already discussed the log-normal model and applied it to the recidivism data, we have not yet considered the shape of its hazard function. Unlike the Weibull model, the log-normal model has a nonmonotonic hazard function. The hazard is 0 when $t=0$. It rises to a peak and then declines toward 0 as t goes to infinity. The log-normal is *not* a proportional hazards model, and its hazard function cannot be expressed in closed form (it involves the c.d.f. of a standard normal variable). It can, however, be expressed as a regression model in which the dependent variable is the logarithm of the hazard. Specifically,

$$\log h(t) = \log h_0(te^{-\beta x}) - \beta x$$

where $h_0(.)$ can be interpreted as the hazard function for an individual with $x = 0$. This equation also applies to the log-logistic and gamma models to be discussed shortly, except that $h_0(.)$ is different in each case.

Some typical log-normal hazard functions are shown in Figure 4.1. All three functions correspond to distributions with a median of 1.0. When σ is large, the hazard peaks so rapidly that the function is almost indistinguishable from those like the Weibull and log-logistic that may have an infinite hazard when $t = 0$.

The inverted U-shape of the log-normal hazard is often appropriate for repeatable events. Suppose, for example, that the event of interest is a residential move. Immediately after a move, the hazard of another move is likely to be extremely low. People need to rest and recoup the substantial costs involved in moving. The hazard will certainly rise with time, but much empirical evidence indicates that it eventually begins to decline. One explanation is that, as time goes by, people become increasingly invested in a particular location or community. However, Chapter 8 shows how the declining portion of the hazard function may also be a consequence of unobserved heterogeneity.

Figure 4.1 *Typical Hazard Functions for a Log-Normal Model*

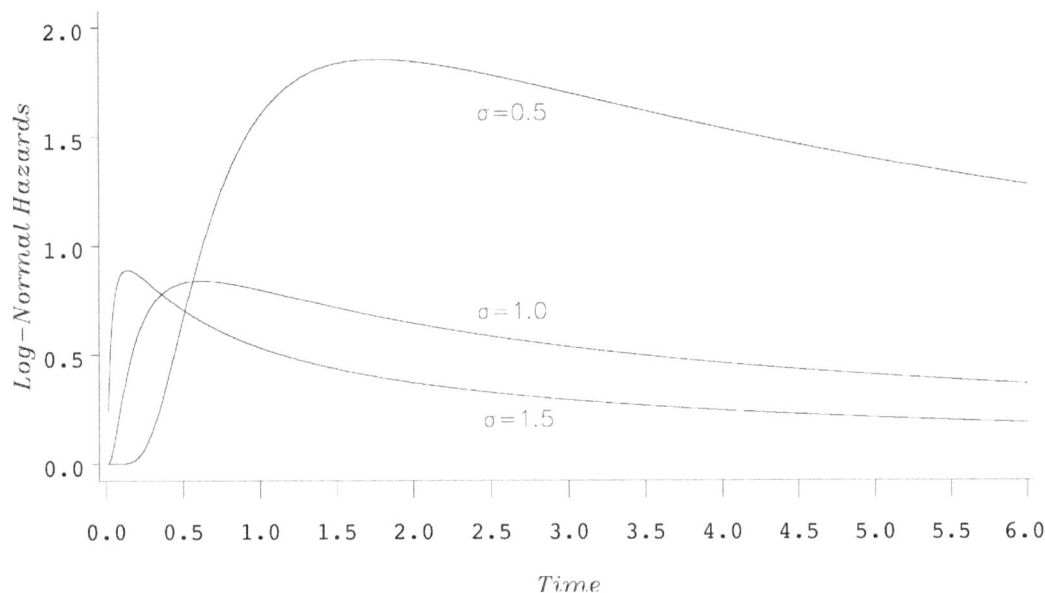

The Log-Logistic Model

Another model that allows for an inverted U-shaped hazard is the log-logistic model, which assumes that ε has a logistic distribution with p.d.f.

$$f(\varepsilon) = \frac{e^{\varepsilon}}{(1 + e^{\varepsilon})^2}.$$

The logistic distribution is symmetric with a mean of 0, and is quite similar in shape to the normal distribution. It is well known to students of the logistic (logit) regression model, which can be derived by assuming a linear model with a logistically distributed error term and a dichotomization of the dependent variable.

If ε has a logistic distribution, then so does log T (although with a nonzero mean). It follows that T has a log-logistic distribution. The log-logistic hazard function is

$$h(t) = \frac{\lambda\gamma(\lambda t)^{\gamma-1}}{1 + (\lambda t)^{\gamma}}$$

where $\gamma = 1/\sigma$ and $\lambda = \exp\{-[\beta_0 + \beta_1 x_1 + \dots + \beta_k x_k]\}$. This produces the characteristic shapes shown in Figure 4.2, all of which correspond to distributions with a median of 1.0.

Figure 4.2 *Typical Hazard Functions for the Log-Logistic Model*

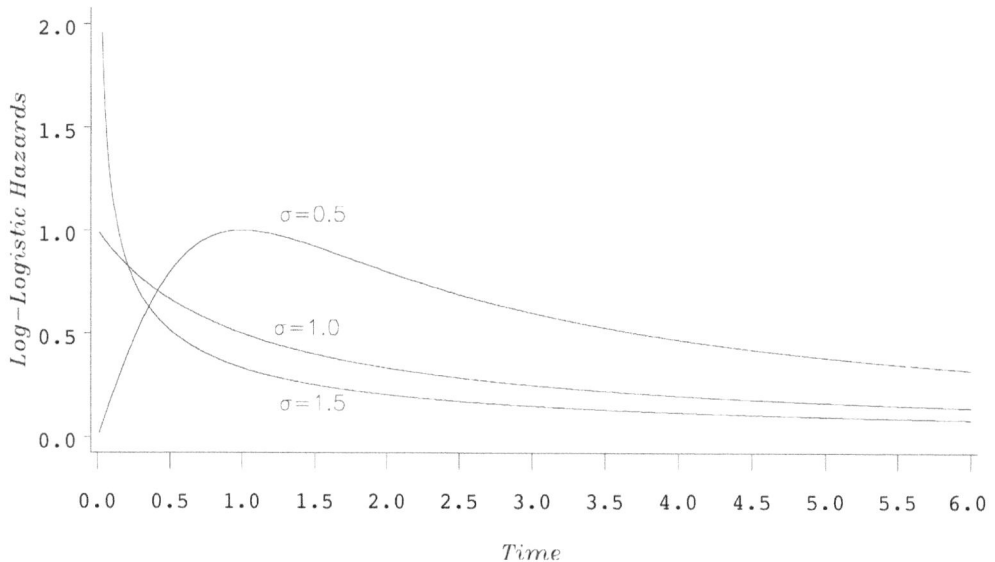

When $\sigma < 1$, the log-logistic hazard is similar to the log-normal hazard: it starts at 0, rises to a peak, and then declines toward 0. When $\sigma > 1$, the hazard behaves like the decreasing Weibull hazard: it starts at infinity and declines toward 0. When $\sigma = 1$, the hazard has a value of λ at $t=0$ and then declines toward 0 as t goes to infinity.

Despite the complexity of its hazard function, the log-logistic model has a rather simple survivor function:

$$S(t) = \frac{1}{1+(\lambda t)^{\gamma}}$$

As before, $\gamma = 1/\sigma$ and $\lambda = \exp\{-[\beta_0 + \beta_1 x_1 + \ldots + \beta_k x_k]\}$. A little algebra shows that this can be written as

$$\log\left[\frac{S(t)}{1-S(t)}\right] = \beta_0^{\bullet} + \beta_1^{\bullet} x_1 + \ldots + \beta_k^{\bullet} x_k - \gamma \log t$$

where $\beta_i^{\bullet} = \beta_i/\sigma$ for $i = 1, \ldots, k$. This is just a logistic regression model for the probability that an event occurs prior to t. Thus, for the recidivism data, the log-logistic model can be estimated by fitting a logistic regression model to the dichotomy arrested versus not arrested in the first year after release. (Because t is a constant 52 weeks, the term $\gamma \log t$ gets absorbed into the intercept.) Of course, this estimation method is not fully efficient because we are not using the information on the exact timing of the arrests, and we certainly will not get the same estimates. The point is that the two apparently different methods are actually estimating the same underlying model.

To fit the log-logistic model with PROC LIFEREG, you specify DISTRIBUTION=LLOGISTIC as an option in the MODEL statement. Output 4.4 shows the results for the recidivism data. The first thing to notice is that the estimate of σ (labeled "Scale") is less than 1.0, implying that the estimated hazard function follows the inverted U-shaped form shown in Figure 4.2. Given what I just said about the similarity of the log-normal and log-logistic hazards, you might expect the other results to be similar to the log-normal output in Output 4.1. But the coefficients and test statistics actually appear to be closer to those for the Weibull model in Output 4.3.

Output 4.4 *Log-Logistic Model Applied to Recidivism Data*

```
             Analysis of Maximum Likelihood Parameter Estimates

                           Standard   95% Confidence    Chi-
Parameter      DF Estimate   Error        Limits       Square Pr > ChiSq

Intercept       1   3.9183   0.4274   3.0805   4.7561   84.03   <.0001
fin             1   0.2889   0.1456   0.0035   0.5742    3.94   0.0472
age             1   0.0364   0.0156   0.0058   0.0669    5.45   0.0195
race            1  -0.2791   0.2297  -0.7293   0.1710    1.48   0.2242
wexp            1   0.1784   0.1572  -0.1297   0.4865    1.29   0.2563
mar             1   0.3473   0.2697  -0.1812   0.8758    1.66   0.1978
paro            1   0.0508   0.1496  -0.2424   0.3440    0.12   0.7341
prio            1  -0.0692   0.0227  -0.1138  -0.0246    9.25   0.0023
Scale           1   0.6471   0.0559   0.5463   0.7666
```

The Gamma Model

The generalized gamma model has one more parameter than any of the other models we have considered, and that fact implies that its hazard function can take on a wide variety of shapes. In particular, the exponential, Weibull, and log-normal models (but not the log-logistic) are all special cases of the generalized gamma model. This fact is exploited later in this chapter when we consider likelihood ratio tests for comparing different models. But the generalized gamma model can also take on shapes that are unlike any of these special cases. Most important, it can have hazard functions with U or *bathtub* shapes in which the hazard declines, reaches a minimum, and then increases. It is well known that the hazard for human mortality, considered over the whole life span, has such a shape. On the other hand, the generalized gamma model cannot represent hazard functions that have more than one reversal of direction.

Given the richness of the generalized gamma model, why not always use it instead of the other models? There are two reasons. First, the formula for the hazard function for the generalized gamma model is rather complicated, involving the gamma function and the incomplete gamma function. Consequently, you may often find it difficult to judge the shape of the hazard function from the estimated parameters. By contrast, hazard functions for the specific submodels can be rather simply described, as we have already seen. Second, computation for the generalized gamma model is more difficult. For example, it took more than five times as much computer time to estimate the generalized gamma model for the recidivism data as compared with the exponential model. This fact can be an important consideration when you are working with very large data sets. The generalized gamma model also has a reputation for convergence problems, although the parameterization and numerical algorithms used by PROC LIFEREG seem to have reduced these to a minimum.

To fit the generalized gamma model with PROC LIFEREG, you specify D=GAMMA as an option in the MODEL statement. Output 4.5 shows the results from fitting this model to the recidivism data. As usual, the Scale parameter is the estimate of σ in equation (4.2). The estimate labeled "Shape" is the additional shape parameter that is denoted by δ in the PROC LIFEREG documentation. (In the output for earlier releases of PROC LIFEREG, this parameter is labeled "Gamma.") When the shape parameter is 0, we get the log-normal distribution. When it is 1.0, we have the Weibull distribution. In Output 4.5, the shape estimate is almost exactly 1.0, so we are very close to the Weibull distribution. Later, we'll make this comparison more rigorous.

Output 4.5 *Generalized Gamma Model Applied to the Recidivism Data*

Parameter	DF	Estimate	Standard Error	95% Confidence Limits		Chi-Square	Pr > ChiSq
Intercept	1	3.9915	0.4349	3.1391	4.8439	84.23	<.0001
fin	1	0.2724	0.1401	-0.0022	0.5471	3.78	0.0518
age	1	0.0407	0.0165	0.0082	0.0731	6.04	0.0140
race	1	-0.2255	0.2280	-0.6723	0.2213	0.98	0.3226
wexp	1	0.1073	0.1659	-0.2179	0.4326	0.42	0.5177
mar	1	0.3118	0.2769	-0.2309	0.8545	1.27	0.2602
paro	1	0.0588	0.1398	-0.2152	0.3328	0.18	0.6741
prio	1	-0.0659	0.0213	-0.1076	-0.0241	9.56	0.0020
Scale	1	0.7151	0.2396	0.3708	1.3790		
Shape	1	0.9943	0.4849	0.0439	1.9446		

CATEGORICAL VARIABLES AND THE CLASS STATEMENT

In the recidivism example, several of the covariates—race, marital status, work experience, and parole status—are dichotomous variables that are coded as indicator (dummy) variables. For categorical covariates with more than two categories, the standard approach is to create a *set* of indicator variables, one for each category (except for one). You can do this in the DATA step, but PROC LIFEREG does it automatically if the variable (or variables) is listed in a CLASS statement. Here's an example. Another covariate in the recidivism data set is education, which was originally coded like this:

2 = 6th grade or less	24 cases
3 = 7th to 9th grade	239 cases
4 = 10th to 11th grade	119 cases
5 = 12th grade	39 cases
6 = some college	11 cases

Due to the small numbers of cases in the two extreme categories, I combined them (in a DATA step) with the adjacent categories to produce a variable EDUC with values of 3 (9th or less), 4 (10th to 11th), and 5 (12th or more). I then specified a Weibull model in PROC LIFEREG with the following statements:

```
PROC LIFEREG DATA=recid;
   CLASS educ;
   MODEL week*arrest(0)=fin age race wexp mar paro
         prio educ / D=WEIBULL;
RUN;
```

Output 4.6 shows the results (some output lines have been omitted). For any variables listed in the CLASS statement, PROC LIFEREG first reports the number of levels found and the values for those levels. The Type III Analysis of Effects table, which really served no purpose in our previous models, now gives us a very useful overall test for EDUC, which (with a *p*-value of .20) is not statistically significant. This test has two degrees of freedom (df), corresponding to the two coefficients that are estimated for EDUC. What is particularly attractive about this test is that it does not depend on which category of EDUC is the omitted category.

In the table of estimates, there are three lines for the EDUC variable. The first two lines contain coefficients, standard errors, and hypothesis tests for levels 3 and 4 of EDUC, while the last line merely informs us that level 5 is the omitted category. Hence, each of the estimated coefficients is a contrast with level 5. (The default in PROC LIFEREG is to take the highest formatted value as the omitted category, but you can get some control over this with the ORDER option in the PROC LIFEREG statement.)

Output 4.6 *Recidivism Model with Education as a CLASS Variable*

```
                        Class Level Information

                  Name        Levels     Values

                  educ          3         3 4 5

                  Type III Analysis of Effects

                                    Wald
          Effect        DF     Chi-Square     Pr > ChiSq

          fin            1        3.8069        0.0510
          age            1        6.0426        0.0140
          race           1        1.2819        0.2575
          wexp           1        0.2581        0.6114
          mar            1        1.2160        0.2701
          paro           1        0.2221        0.6374
          prio           1        7.5308        0.0061
          educ           2        3.2023        0.2017

          Analysis of Maximum Likelihood Parameter Estimates
```

Parameter		DF	Estimate	Standard Error	95% Confidence Limits		Chi-Square	Pr > ChiSq
Intercept		1	4.4680	0.5171	3.4544	5.4816	74.65	<.0001
fin		1	0.2690	0.1379	-0.0012	0.5392	3.81	0.0510
age		1	0.0392	0.0159	0.0079	0.0705	6.04	0.0140
race		1	-0.2524	0.2229	-0.6893	0.1845	1.28	0.2575
wexp		1	0.0773	0.1522	-0.2209	0.3755	0.26	0.6114
mar		1	0.3013	0.2732	-0.2342	0.8368	1.22	0.2701
paro		1	0.0658	0.1396	-0.2078	0.3394	0.22	0.6374
prio		1	-0.0585	0.0213	-0.1004	-0.0167	7.53	0.0061
educ	3	1	-0.5116	0.3090	-1.1172	0.0941	2.74	0.0978
educ	4	1	-0.3536	0.3243	-0.9892	0.2819	1.19	0.2755
educ	5	0	0.0000	.	.	.	.	.
Scale		1	0.7119	0.0634	0.5979	0.8476		
Weibull Shape		1	1.4047	0.1251	1.1798	1.6726		

Unlike most other SAS procedures that have a CLASS statement, PROC LIFEREG does not have a TEST or CONTRAST statement that would enable you to test whether EDUC 3 has the same coefficient as

EDUC 4. Methods for constructing such a test by hand calculation are described in the **Hypothesis Tests** section below.

In earlier releases of PROC LIFEREG, it was not possible to include interaction terms directly in the MODEL statement. SAS 9.2 allows interactions for any set of variables using the familiar notation X1*X2.

MAXIMUM LIKELIHOOD ESTIMATION

All models in PROC LIFEREG are estimated by the method of maximum likelihood. This section explores some of the basics of ML estimation, with an emphasis on how it handles censored observations. The discussion is not intended to be rigorous. If you want a more complete and careful treatment of ML, you should consult one of the many texts available on the subject. For example, Kalbfleisch and Prentice (2002) give a more detailed introduction in the context of survival analysis.

ML is a quite general approach to estimation that has become popular in many different areas of application. There are two reasons for this popularity. First, ML produces estimators that have good large-sample properties. Provided that certain regularity conditions are met, ML estimators are consistent, asymptotically efficient, and asymptotically normal. *Consistency* means that the estimates converge *in probability* to the true values as the sample gets larger, implying that the estimates will be approximately unbiased in large samples. *Asymptotically efficient* means that, in large samples, the estimates will have standard errors that are (approximately) at least as small as those for any other estimation method. And, finally, *asymptotically normal* means that the sampling distribution of the estimates will be approximately normal in large samples, which implies that you can use the normal and chi-square distributions to compute confidence intervals and *p*-values.

All these approximations get better as the sample size gets larger. The fact that these desirable properties have been proven only for large samples does *not* mean that ML has bad properties for small samples. It simply means that we often don't know what the small-sample properties are. And in the absence of attractive alternatives, researchers routinely use ML estimation for both large and small samples. Although I won't argue against that practice, I do urge caution in interpreting *p*-values and confidence intervals when samples are small. Despite the temptation to accept *larger p*-values as evidence against the null hypothesis in small samples, it is actually more reasonable to demand *smaller* values to compensate for the fact that the approximation to the normal or chi-square distributions may be poor.

The other reason for ML's popularity is that it is often straightforward to derive ML estimators when there are no other obvious possibilities. As we will see, one case that ML handles nicely is data with censored observations. Although you can use least squares with certain adjustments for censoring (Lawless, 2002), such estimates often have much larger standard errors, and there is little available theory to justify the construction of hypothesis tests or confidence intervals.

The basic principle of ML is to choose as estimates those values that will maximize the probability of observing what we have, in fact, observed. There are two steps to this: (1) write down a formula for the probability of the data as a function of the unknown parameters, and (2) find the values of the unknown parameters that make the value of this formula as large as possible.

The first step is known as *constructing the likelihood function*. To accomplish this, you must specify a model, which amounts to choosing a probability distribution for the dependent variable and choosing a functional form that relates the parameters of this distribution to the values of the covariates. We have already considered those two choices. The second step—*maximization*—typically requires an iterative numerical method that involves successive approximations.

In the next section, I work through the basic mathematics of constructing and maximizing the likelihood function. You can skip this part without loss of continuity if you're not interested in the details or if you simply want to postpone the effort. Immediately after this section, I discuss some of the practical details of ML estimation with PROC LIFEREG.

Maximum Likelihood Estimation: Mathematics

Assume that we have n independent individuals ($i = 1, ..., n$). For each individual i, the data consist of three parts: t_i, δ_i, and $\mathbf{x}_i$, where t_i is the time of the event or the time of censoring, δ_i, is an indicator variable with a value of 1 if t_i is uncensored or 0 if right censored, and $\mathbf{x}_i = [1 \ x_{i1} \ ... \ x_{ik}]'$ is a vector of covariate values (the 1 is for the intercept). For simplicity, we treat $\mathbf{x}_i$ as fixed rather than random. We could get equivalent results if $\mathbf{x}_i$ were random and the distributions of δ_i and t_i were expressed conditional on the values of $\mathbf{x}_i$. But that would just complicate the notation. We also assume that censoring is non-informative.

For the moment, suppose that all the observations are uncensored. Because we are assuming independence, it follows that the probability of the entire data is found by taking the product of the probabilities of the data for every individual. Because t_i is assumed to be measured on a continuum, the probability that it will take on any specific value is 0.

Instead, we represent the probability of each observation by the probability density function (p.d.f.), $f(t_i)$. Thus, the probability (or likelihood) of the data is given by the following expression, where Π indicates repeated multiplication:

$$L = \prod_{i=1}^{n} f_i(t_i).$$

Notice that f_i is subscripted to indicate that each individual has a different p.d.f. that depends on the covariates.

To proceed further, we need to substitute an expression for $f_i(t_i)$ that involves the covariates and the unknown parameters. Before we do that, however, let's see how this likelihood is altered if we have censored cases. If an individual is censored at time t_i, all we know is that the individual's event time is greater than t_i. But the probability of an event time greater than t_i is given by the survivor function $S(t)$ evaluated at time t_i. Now suppose that we have r uncensored observations and $n - r$ censored observations. If we arrange the data so that all the uncensored cases come first, we can write the likelihood as

$$L = \prod_{i=1}^{r} f_i(t_i) \prod_{i=r+1}^{n} S_i(t_i)$$

where, again, we subscript the survivor function to indicate that it depends on the covariates. Using the censoring indicator δ, we can equivalently write this as

$$L = \prod_{i=1}^{n} [f_i(t_i)]^{\delta_i} [S_i(t_i)]^{1-\delta_i}$$

Here δ_i acts as a switch, turning the appropriate function on or off, depending on whether the observation is censored. As a result, we do not need to order the observations by censoring status. This last expression, which applies to all the models that PROC LIFEREG estimates with right-censored data, shows how censored and uncensored cases are combined in ML estimation.

Once we choose a particular model, we can substitute appropriate expressions for the p.d.f. and the survivor function. Let's take the simplest case—the exponential model. We have

$$f_i(t_i) = \lambda_i e^{-\lambda_i t_i} \text{ and } S_i(t_i) = e^{-\lambda_i t_i}$$

where $\lambda_i = \exp(-\boldsymbol{\beta} \mathbf{x}_i)$ and $\boldsymbol{\beta}$ is a vector of coefficients. Substituting, we get

$$L = \prod_{i=1}^{n} \left[\lambda_i e^{-\lambda_i t_i} \right]^{\delta_i} \left[e^{-\lambda_i t_i} \right]^{1-\delta_i} = \prod_{i=1}^{n} \lambda_i^{\delta_i} e^{-\lambda_i t_i}.$$

Although this expression can be maximized directly, it is generally easier to work with the natural logarithm of the likelihood function because products get converted into sums and exponents become coefficients. Because the logarithm is an increasing function, whatever maximizes the logarithm also maximizes the original function.

Taking the logarithm of the likelihood, we get

$$\log L = \sum_{i=1}^{n} \delta_i \log \lambda_i - \sum_{i=1}^{n} \lambda_i t_i$$

$$= -\beta \sum_{i=1}^{n} \delta_i \mathbf{x}_i - \sum_{i=1}^{n} t_i e^{-\beta \mathbf{x}_i}$$

Now we are ready for step 2, finding values of β that make this expression as large as possible. There are many different methods for maximizing functions like this. One well-known approach is to find the derivative of the function with respect to β, set the derivative equal to 0, and then solve for β. Taking the derivative and setting it equal to 0 gives us

$$\sum_{i=1}^{n} \delta_i \mathbf{x}_i = \sum_{i=1}^{n} \mathbf{x}_i t_i e^{-\beta \mathbf{x}_i} .$$

Because $\mathbf{x}_i$ is a vector, this is actually a system of $k + 1$ equations, one for each element of β. While these equations are not terribly complicated, the problem is that they involve nonlinear functions of β. Consequently, except in special cases (like a single dichotomous x variable), there is no explicit solution. Instead, we have to rely on iterative methods, which amount to successive approximations to the solution until the approximations converge to the correct value. Again, there are many different methods for doing this. All give the same solution, but they differ in such factors as speed of convergence, sensitivity to starting values, and computational difficulty at each iteration.

PROC LIFEREG uses the Newton-Raphson algorithm (actually a *ridge stabilized* version of the algorithm), which is by far the most popular numerical method for solving for β. The method is named after Sir Isaac Newton, who devised it for a single equation and a single unknown. But who was Raphson? Joseph Raphson was a younger contemporary of Newton who generalized the algorithm to multiple equations with multiple unknowns.

The Newton-Raphson algorithm can be described as follows. Let $\mathbf{U}(\beta)$ be the vector of first derivatives of $\log L$ with respect to β and let $\mathbf{I}(\beta)$ be the matrix of second derivatives of $\log L$ with respect to β. That is,

$$U(\beta) = \frac{\partial \log L}{\partial \beta} = -\sum_{i=1}^{n} \delta_i \mathbf{x}_i + \sum_{i=1}^{n} t_i \exp(-\beta \mathbf{x}_i) \mathbf{x}_i$$

$$I(\beta) = \frac{\partial^2 \log L}{\partial \beta \partial \beta'} = -\sum_{i=1}^{n} t_i \exp(-\beta \mathbf{x}_i) \mathbf{x}_i \mathbf{x}_i'$$

The vector of first derivatives $U(\beta)$ is sometimes called the *gradient* or *score*, while the matrix of second derivatives $I(\beta)$ is called the *Hessian*. The Newton-Raphson algorithm is then

$$\beta_{j+1} = \beta_j - I^{-1}(\beta_j)U(\beta_j) \tag{4.4}$$

where I^{-1} is the inverse of I. In practice, we need a set of starting values β_0, which PROC LIFEREG calculates by using ordinary least squares, treating the censored observations as though they were uncensored. These starting values are substituted into the right side of equation (4.4), which yields the result for the first iteration, β_1. These values are then substituted back into the right side, the first and second derivatives are recomputed, and the result is β_2. This process is repeated until the maximum change in the parameter estimates from one step to the next is less than .00000001. (This is an absolute change if the current parameter value is less than .01; otherwise, it is a relative change.)

Once the solution is found, a convenient by-product of the Newton-Raphson algorithm is an estimate of the covariance matrix of the coefficients, which is just $-I^{-1}(\hat{\beta}_j)$. This matrix, which can be printed by listing COVB as an option in the MODEL statement, is often useful for constructing hypothesis tests about linear combinations of coefficients. PROC LIFEREG computes standard errors of the parameters by taking the square roots of the main diagonal elements of this matrix.

Maximum Likelihood Estimation: Practical Details

PROC LIFEREG chooses parameter estimates that maximize the logarithm of the likelihood of the data. For the most part, the iterative methods used to accomplish this task work quite well with no attention from the data analyst. If you're curious to see how the iterative process works, you can request ITPRINT as an option in the MODEL statement. Then, for each iteration, PROC LIFEREG will print out the log-likelihood and the parameter estimates. When the iterations are complete, the final gradient vector and the negative of the Hessian matrix will also be printed (see the preceding section for definitions of these quantities).

When the exponential model was fitted to the recidivism data, the ITPRINT output revealed that it took six iterations to reach a solution. The log-likelihood for the starting values was −531.1, which increased to

−327.5 at convergence. Examination of the coefficient estimates showed only slight changes after the fourth iteration. By comparison, the generalized gamma model took 13 iterations to converge.

Occasionally the algorithm fails to converge, although this seems to occur much less frequently than it does with logistic regression. In general, nonconvergence is more likely to occur when samples are small, when censoring is heavy, or when many parameters are being estimated. There is one situation, in particular, that guarantees nonconvergence (at least in principle). If all the cases at one value of a dichotomous covariate are censored, the coefficient for that variable becomes larger in magnitude at each iteration. Here's why: the coefficient of a dichotomous covariate is a function of the logarithm of the ratio of the hazards for the two groups. But if all the cases in a group are censored, the ML estimate for the hazard in that group is 0. If the 0 is in the denominator of the ratio, then the coefficient tends toward plus infinity. If it's in the numerator, taking the logarithm yields a result that tends toward minus infinity. By extension, if a covariate has multiple values that are treated as a set of dichotomous variables (for example, with a CLASS statement) and all cases are censored for one or more of the values, nonconvergence should result. When this happens, there is no ideal solution. You can remove the offending variable from the model, but that variable may actually be one of the strongest predictors. When the variable has more than two values, you can combine adjacent values or treat the variable as quantitative.

PROC LIFEREG has two ways of alerting you to convergence problems. If the number of iterations exceeds the maximum allowed (the default is 50), SAS issues the message: WARNING: Convergence not attained in 50 iterations. WARNING: The procedure is continuing but the validity of the model fit is questionable. If it detects a problem before the iteration limit is reached, the software says WARNING: The negative of the Hessian is not positive definite. The convergence is questionable. Unfortunately, PROC LIFEREG sometimes reports estimates and gives no warning message in situations that are fundamentally nonconvergent. The only indication of a problem is a coefficient that is large in magnitude together with a huge standard error.

It's tempting to try to get convergence by raising the default maximum number of iterations or by relaxing the convergence criterion. This rarely works, however, so don't get your hopes up. You can raise the maximum with the MAXITER= option in the MODEL statement. You can alter the convergence criterion with the CONVERGE= option, but I don't recommend this unless you know what you're doing. Too large a value could make it seem that convergence had occurred when there is actually no ML solution.

HYPOTHESIS TESTS

PROC LIFEREG is somewhat skimpy in its facilities for hypothesis tests. It automatically reports a chi-square test for the hypothesis that each coefficient is 0. These are Wald tests that are calculated simply by dividing each coefficient by its estimated standard error and squaring the result. For models like the exponential that restrict the scale parameter to 1.0, PROC LIFEREG reports a Lagrange multiplier chi-square statistic (also known as a score statistic) for the hypothesis that the parameter is, indeed, equal to 1.0. Finally, as we've seen, for categorical variables named in a CLASS statement, PROC LIFEREG gives a Wald chi-square statistic for the null hypothesis that all the coefficients associated with the variable are 0.

To test other hypotheses, you have to construct the appropriate statistic yourself. Before describing how to do this, I'll first present some background. For all the regression models considered in this book, there are three general methods for constructing test statistics: Wald statistics, score statistics, and likelihood-ratio statistics. Wald statistics are calculated using certain functions (quadratic forms) of parameter estimates and their estimated variances and covariances. Score statistics are based on similar functions of the first and second derivatives of the log-likelihood function. Finally, likelihood-ratio statistics are calculated by maximizing the likelihood twice: under the null hypothesis and with the null hypothesis relaxed. The statistic is then twice the positive difference in the two log-likelihoods.

You can use all three methods to test the same hypotheses, and all three produce chi-square statistics with the same number of degrees of freedom. Furthermore, they are asymptotically equivalent, meaning that their approximate large-sample distributions are identical. Hence, asymptotic theory gives no basis for preferring one method over another. There is some evidence that likelihood-ratio statistics may more closely approximate a chi-square distribution in small- to moderate-sized samples, however, and some authors (for example, Collett, 2002) express a strong preference for these statistics. On the other hand, Wald tests and score tests are often more convenient to calculate because they don't require re-estimation of the model for each hypothesis tested.

Let's first consider a likelihood-ratio test of the null hypothesis that all the covariates have coefficients of 0. This is analogous to the usual *F*-test that is routinely reported for linear regression models. (Many authorities hold that if this hypothesis is not rejected, then there is no point in examining individual coefficients for statistical significance.) To calculate this statistic, we need only to fit a null model that includes no covariates. For a Weibull model, we can accomplish that with the following statement:

```
MODEL week*arrest(0)= / D=WEIBULL;
```

For the recidivism data, this produces a log-likelihood of -338.59. By contrast, the Weibull model with seven covariates displayed in Output 4.3 has a log-likelihood of -321.85. Taking twice the positive difference between these two values yields a chi-square value of 33.48. With seven degrees of freedom (the number of covariates excluded from the null model), the *p*-value is less than .001. So we reject the null hypothesis and conclude that at least one of the coefficients is nonzero.

You can also test the same hypothesis with a Wald statistic, but that involves the following steps:

1. request that the parameter estimates and their covariance matrix be written to a SAS data set
2. read that data set into PROC IML, the SAS matrix algebra procedure
3. use PROC IML to perform the necessary matrix calculations.

(These calculations include inverting the appropriate submatrix of the covariance matrix and premultiplying and postmultiplying that matrix by a vector containing appropriate linear combinations of the coefficients.) That's clearly a much more involved procedure.

Wald statistics for testing the equality of any two coefficients *are* simple to calculate. The method is particularly useful for doing post-hoc comparisons of the coefficients of CLASS variables. Earlier we used a CLASS statement to include a three-category education variable in the model. As shown in Output 4.6, there is one chi-square test comparing category 3 with category 5 and another chi-square test comparing category 4 with category 5. But there is no test reported for comparing category 3 with category 4. The appropriate null hypothesis is that $\beta_3 = \beta_4$, where the subscripts refer to the values of categories. A Wald chi-square for testing this hypothesis can be computed by

$$\frac{(\hat{\beta}_3 - \hat{\beta}_4)^2}{Var(\hat{\beta}_3) + Var(\hat{\beta}_4) - 2Cov(\hat{\beta}_3, \hat{\beta}_4)}. \tag{4.5}$$

Estimates of the variances and covariances in the denominator are easily obtained from the covariance matrix that was requested in the MODEL statement. Output 4.7 shows a portion of the printed matrix.

Output 4.7 *A Portion of the Covariance Matrix for a Model with a CLASS Variable*

	educ3	educ4	Scale
Intercept	-0.094637	-0.090901	0.003377
fin	-0.000920	-0.001186	0.001338
age	0.000137	0.000285	0.000202
race	0.002250	-0.003371	-0.001341
wexp	0.002200	-0.001100	0.000398
mar	0.000551	0.000826	0.001593
paro	-0.000617	0.000704	0.000383
prio	-0.000593	-0.000131	-0.000260
educ3	0.095495	0.086663	-0.002738
educ4	0.086663	0.105149	-0.001756

Estimated Covariance Matrix

$Var\left(\hat{\beta}_3\right)$ is found to be .095495 at the intersection of educ3 with itself, and similarly $Var\left(\hat{\beta}_4\right)$ is .105149 at the intersection of educ4 with itself. The covariance is .086663 at the intersection of educ3 and educ4. Combining these numbers with the coefficient estimates in Output 4.7, we get

$$\frac{\left[.5116 - (-.3536)\right]^2}{.09549 + .1051 - 2(.08666)} = .9154 .$$

With 1 degree of freedom, the chi-square value is far from the .05 critical value of 3.84. We conclude that there is no difference in the time to arrest between those with 9th grade or less and those with 10th or 11th grade education. This should not be surprising because the overall chi-square test is not significant, nor is the more extreme comparison of category 3 with category 5. Of course, another way to get this same test statistic is simply to re-run the model with category 4 as the omitted category rather than category 5 (which can be accomplished by recoding the variable so that category 4 has the highest value). The chi-square statistic for category 3 will then be equal to .9154.

When performing post-hoc comparisons like this, it is generally advisable to adjust the alpha level for multiple comparisons. The simplest approach is the well-known Bonferroni method: For k tests and an overall Type I error rate of α, each test uses α/k as the criterion value.

We can also test the hypothesis that $\beta_3 = \beta_4$ with the likelihood ratio statistic. To do that, we must re-estimate the model while imposing the constraint that $\beta_3 = \beta_4$. We can do this by recoding the education variable so that levels 3 and 4 have the same value. For example, the DATA step can contain a statement like

```
IF educ = 3 THEN educ = 4;
```

When I estimated the model with this recoding, the log-likelihood was −317.97, compared with −317.50 with the original coding. Twice the positive difference is .94, which, again, is far from statistically significant.

GOODNESS-OF-FIT TESTS WITH THE LIKELIHOOD-RATIO STATISTIC

As we have seen, the AFT model encompasses a number of submodels that differ in the assumed distribution for T, the time of the event. When we tried out those models on the recidivism data, we found that they produced generally similar coefficient estimates and p-values. A glaring exception is the log-normal model, which yields qualitatively different conclusions for some of the covariates. Clearly, we need some way of deciding between the log-normal and the other models. Even if all the models agree on the coefficient estimates, they still have markedly different implications for the shape of the hazard function. Again we may need methods for deciding which of these shapes is the best description of the true hazard function.

In the next section, we'll consider some graphical methods for comparing models. Here, we examine a simple and often decisive method based on the likelihood-ratio statistic. In general, likelihood-ratio statistics can be used to compare nested models. A model is said to be nested within another model if the first model is a special case of the second. More precisely, model A is nested within model B if A can be obtained by imposing restrictions on the parameters in B. For example, the exponential model is nested within the Weibull model. You get the exponential from the Weibull by forcing the scale parameter σ equal to 1, and you get the exponential from the gamma by forcing both the shape and scale parameters equal to 1.

If model A is nested within model B, we can evaluate the fit of A by taking twice the positive difference in the log-likelihoods for the two models. Of course, to evaluate a model in this way, you need to find another model within which it is nested. As previously noted, the Weibull and log-normal models (but not the log-logistic) are both nested within the generalized gamma model, making it a simple matter to evaluate them with the likelihood-ratio test.

Here are the restrictions on the generalized gamma that are implied by its submodels:

$\delta = 1$	Weibull
$\sigma = 1, \delta = 1$	exponential
$\delta = 0$	log-normal

Remember that σ is the scale parameter and δ is the shape parameter. The likelihood-ratio test for each of these models is, in essence, a test for the null hypothesis that the particular restriction is true. Hence, these tests should be viewed not as omnibus tests of the fit of a model but rather as tests of particular features of a model and its fit to the data.

Let's calculate the likelihood-ratio tests for the recidivism data. The log-likelihoods for the models fitted earlier in this chapter are

−325.83	exponential
−319.38	Weibull
−322.69	log-normal
−319.40	log-logistic
−319.38	generalized gamma

Because these log-likelihoods are all negative (which will virtually always be the case), lower magnitudes correspond to better fits. Taking the differences between nested models and multiplying by 2 yields the following likelihood-ratio chi-square statistics:

12.90	exponential vs. Weibull
12.90	exponential vs. g. gamma
.00	Weibull vs. g. gamma
6.62	log-normal vs. g. gamma

With the exception of the exponential model versus the generalized gamma model (which has 2 d.f.), all these tests have a single degree of freedom corresponding to the single restriction being tested.

The conclusions are clear. The exponential model must be rejected ($p=.002$), implying that the hazard of arrest is not constant over the 1-year interval. This is consistent with the results we saw earlier for the Lagrange multiplier test, although that test produced a chi-square value of 24.93, more than twice as large as the likelihood-ratio statistic. The log-normal model must also be rejected, although somewhat less decisively ($p=.01$). On the other hand, the Weibull model fits the data very well. Apparently, we can safely disregard the discrepant coefficient estimates for the log-normal model because the model is not consistent with the data.

Before embracing the Weibull model, however, remember that the log-logistic model, which has a nonmonotonic hazard function, does not fit into our nesting scheme. For the recidivism data, its log-likelihood is only trivially lower than that for the gamma model, suggesting a very good fit to the data. While this fact should lead us to retain the log-logistic model as one of our possible candidates, we cannot use it in a formal test of significance.

In interpreting these likelihood-ratio statistics, you should keep in mind that the validity of each test rests on the (at least approximate) truth of the more general model. If that model does not fit the data well, then the test can be quite misleading. I have seen several examples in which the test for the exponential model versus the Weibull model is not significant, but the test for the Weibull model versus the generalized gamma model is highly significant. Without seeing the second test, you might conclude that the hazard is constant when, in fact, it is not. But what about the generalized gamma model itself? How do we know that it provides a decent fit to the data? Unfortunately, we can't get a likelihood-ratio test of the generalized gamma model unless we can fit an even *more* general model. And even if we could fit a more general model, how would we know *that* model was satisfactory? Obviously you have to stop somewhere. As noted earlier, the generalized gamma model is a rich family of distributions, so we expect it to provide a reasonably good fit in the majority of cases.

GRAPHICAL METHODS FOR EVALUATING MODEL FIT

Another way to discriminate between different probability distributions is to use graphical diagnostics. In Chapter 3, we saw how to use PROC LIFETEST to get plots of the estimated survivor function, which can be used to evaluate two of the distributional models considered in this chapter. Specifically, the PLOTS=LS option produces a plot of $-\log \hat{S}(t)$ versus t. If the true distribution is exponential, this plot should yield a straight line with an origin at 0. The LLS option produces a plot of $\log[-\log \hat{S}(t)]$ versus $\log t$, which should be a straight line if the true distribution is Weibull.

There are two limitations to these methods. First, PROC LIFETEST cannot produce any graphs suitable for evaluation of the gamma, log-normal or log-logistic. Second, these graphs do not adjust for the effects of covariates. Both limitations are removed by the PROBPLOT statement in PROC LIFEREG.

The PROBPLOT statement could hardly be easier to use. After the MODEL statement, simply write

```
PROBPLOT;
```

Although several options are available, they are usually not necessary.

The PROBPLOT statement produces non-parametric estimates of the survivor function using a modified Kaplan-Meier method that adjusts for covariates. It then applies a transformation to the survivor estimates that, when graphed against the log of time, should appear as a straight line if the specified model is correct.

Output 4.8 displays a graph produced by the PROBPLOT statement for the exponential model applied to the recidivism data. The upward sloping straight line represents the survival function predicted by the model. The shaded bands around that line are the 95% confidence bands. The circles are the non-parametric survival function estimates. Ideally, all the non-parametric estimates should lie within the confidence bands.

Output 4.8 *Probability Plot for Exponential Model Applied to Recidivism Data*

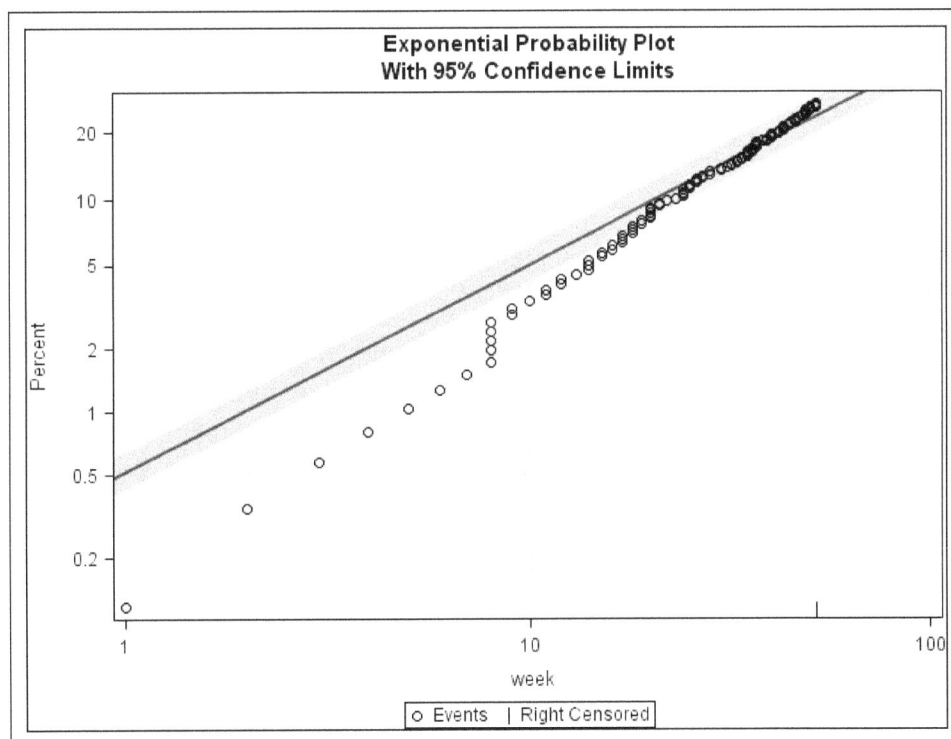

Output 4.9 displays the probability plot for the log-normal model. Again, we see evidence that the model doesn't fit the data well, although it's not nearly as bad as the exponential model. Finally, Output 4.10 shows the probability plot for the Weibull model. Here all the non-parametric estimates fall within the 95% confidence bands.

Output 4.9 *Probability Plot for Log-Normal Model Applied to Recidivism Data*

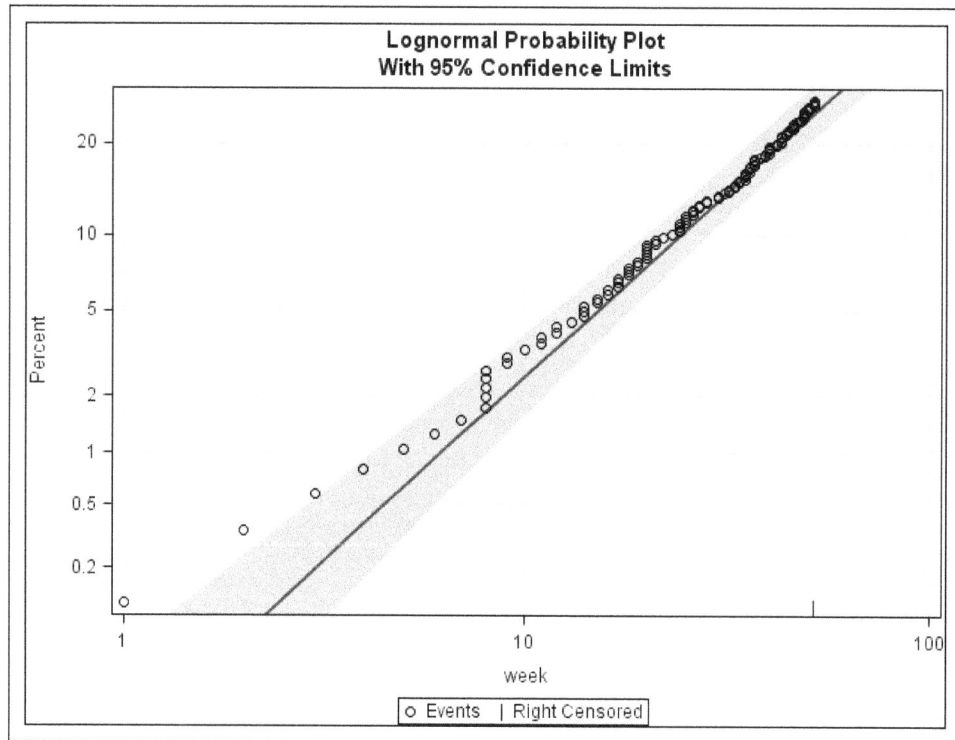

The probability plots for these three models are consistent with the likelihood-ratio tests reported earlier. The Weibull model fits well but the exponential and log-normal models do not. Such consistency will not always occur, however. Unlike the likelihood-ratio tests, which compare one model with another, the probability plots compare the model against the data and thus allow us to evaluate the overall goodness of fit of the model.

Output 4.10 *Probability Plot for Weibull Model Applied to Recidivism Data*

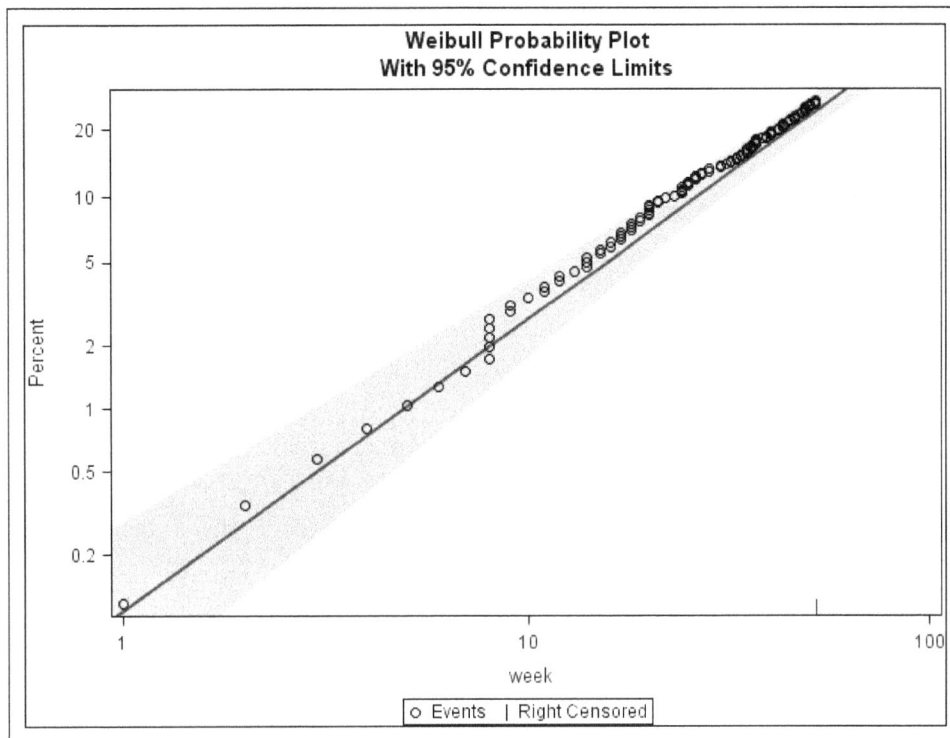

LEFT CENSORING AND INTERVAL CENSORING

One of PROC LIFEREG's more useful features is its ability to handle left censoring and interval censoring. Recall that left censoring occurs when we know that an event occurred earlier than some time t, but we don't know exactly when. Interval censoring occurs when the time of event occurrence is known to be somewhere between times a and b, but we don't know exactly when. Left censoring can be seen as a special case of interval censoring in which $a = 0$; right censoring is a special case in which $b=\infty$.

Interval-censored data are readily incorporated into the likelihood function. The contribution to the likelihood for an observation censored between a and b is just $S_i(a) - S_i(b)$, where $S_i(.)$ is the survivor function for observation i. (This difference is always greater than or equal to 0 because $S_i(t)$ is a nonincreasing function of t.) In other words, the probability of an event occurring in the interval (a, b) is the probability of an event occurring after a minus the probability of it occurring after b. For left-censored data, $S_i(a) = 1$; for right-censored data, $S_i(b)=0$.

PROC LIFEREG can handle any combination of left-censored, right-censored, and interval-censored data, but a different MODEL statement syntax is required if there are any left-censored or interval-censored observations. Instead of a time variable and a censoring variable, PROC LIFEREG needs two time variables, an upper time and a lower time. Let's call them UPPER and LOWER. The MODEL statement then reads as follows:

```
MODEL (lower,upper)=list of covariates;
```

The censoring status is determined by whether the two values are equal and whether either is coded as missing data:

Uncensored:	LOWER and UPPER are both present and equal.
Interval Censored:	LOWER and UPPER are present and different.
Right Censored:	LOWER is present, but UPPER is missing.
Left Censored:	LOWER is missing, but UPPER is present.

You might think that left censoring could also be indicated by coding LOWER as 0, but PROC LIFEREG *excludes* any observations with times that are 0 or negative. Observations are also excluded if both UPPER and LOWER are missing or if LOWER > UPPER. Here are some examples:

Observation	Lower	Upper	Status
1	3.9	3.9	Uncensored
2	7.2	.	Right Censored
3	4.1	5.6	Interval Censored
4	.	2.0	Left Censored
5	0	5.8	Excluded
6	3.2	1.9	Excluded

Let's look at an example of left censoring for the recidivism data. Suppose that of the 114 arrests, the week of arrest was unknown for 30 cases. In other words, we know that an arrest occurred between 0 and 52 weeks, but we don't know when. To illustrate this, I modified the recidivism data by recoding the WEEK variable as missing for the first 30 arrests in the data set. This was accomplished with the following program:

```
PROC SORT DATA=recid OUT=recid2;
  BY DESCENDING arrest;
DATA recidlft;
  SET recid2;
  IF _N_ LE 30 THEN week = .;
RUN;
```

The following program further modifies the data set to create the UPPER and LOWER variables needed for PROC LIFEREG:

```
DATA recid3;
   SET recidlft;
       /* uncensored cases: */
   IF arrest=1 AND week ne . THEN DO;
       upper=week;
       lower=week;
   END;
       /* left-censored cases: */
   IF arrest=1 AND week = . THEN DO;
       upper=52;
       lower=.;
   END;
       /* right-censored cases: */
   IF arrest=0 THEN DO;
       upper=.;
       lower=52;
   END;
RUN;
```

The code for estimating a Weibull model is then

```
PROC LIFEREG DATA=recid3;
   MODEL (lower,upper)=fin age race wexp mar paro prio
           / D=WEIBULL;
RUN;
```

Results in Output 4.11 should be compared with those in Output 4.3, for which there were no left-censored cases. Although the results are quite similar, the chi-square statistics are nearly all smaller when some of the data are left censored. This is to be expected because left censoring entails some loss of information. Note that you cannot compare the log-likelihood for this model with the log-likelihood for the model with no left censoring. Whenever you alter the data, the log-likelihoods are no longer comparable.

Output 4.11 *Results for the Weibull Model with Left-Censored Data*

```
                        Model Information

              Data Set                   WORK.DATA1
              Dependent Variable         Log(lower)
              Dependent Variable         Log(upper)
              Number of Observations            432
              Noncensored Values                 84
              Right Censored Values             318
              Left Censored Values               30
              Interval Censored Values            0
              Name of Distribution          Weibull
              Log Likelihood          -294.0611169

         Analysis of Maximum Likelihood Parameter Estimates

                          Standard   95% Confidence    Chi-
   Parameter    DF Estimate   Error       Limits      Square Pr > ChiSq

   Intercept     1   3.9567  0.4078   3.1574   4.7560  94.14   <.0001
   fin           1   0.2490  0.1348  -0.0152   0.5133   3.41   0.0647
   age           1   0.0413  0.0158   0.0102   0.0723   6.79   0.0092
   race          1  -0.2191  0.2145  -0.6395   0.2013   1.04   0.3070
   wexp          1   0.0777  0.1484  -0.2132   0.3686   0.27   0.6006
   mar           1   0.3001  0.2658  -0.2208   0.8211   1.27   0.2588
   paro          1   0.0682  0.1356  -0.1975   0.3339   0.25   0.6151
   prio          1  -0.0623  0.0207  -0.1030  -0.0217   9.03   0.0027
   Scale         1   0.6911  0.0717   0.5639   0.8470
   Weibull Shape 1   1.4469  0.1501   1.1806   1.7732
```

The recidivism data can also be used to illustrate *interval* censoring. Because we know only the week of the arrest and not the exact day, we can actually view each arrest time as an interval-censored observation. Although Petersen (1991) has shown that some bias can result from treating discrete data as continuous, there's probably little danger of bias with 52 different values for the measurement of arrest time. Nonetheless, we can use the interval-censoring option to get a slightly improved estimate. For an arrest that occurs in week 2, the actual interval in which the arrest occurred is (1, 2). Similarly, the interval is (2, 3) for an arrest occurring in week 3. This suggests the following recoding of the data:

```
DATA recidint;
   SET recid;
      /* interval-censored cases: */
   IF arrest=1 THEN DO;
      upper=week;
      lower=week-.9999;
   END;
      /* right-censored cases: */
   IF arrest=0 THEN DO;
      upper=.;
      lower=52;
   END;
RUN;
PROC LIFEREG DATA=recidint;
   MODEL (lower, upper) = fin age race wexp mar paro prio
         / D=WEIBULL;
RUN;
```

To get the lower value for the interval-censored cases, I subtracted .9999 instead of 1 so that the result is not 0 for those persons with WEEK=1 (which would cause PROC LIFEREG to exclude the observation).

The results in Output 4.12 are very close to those in Output 4.3, which assumed that time was measured exactly. (Some output lines are deleted.) If the intervals had been larger, we might have found more substantial differences. The magnitude of the log-likelihood is nearly doubled for the interval-censored version but, again, log-likelihoods are not comparable when the data are altered.

Output 4.12 *Results Treating Recidivism Data as Interval Censored*

```
                       Model Information

          Data Set                    WORK.RECIDINT
          Dependent Variable           Log(lower)
          Dependent Variable           Log(upper)
          Number of Observations            432
          Noncensored Values                  0
          Right Censored Values             318
          Left Censored Values                0
          Interval Censored Values          114
          Name of Distribution          Weibull
          Log Likelihood             -680.995873

       Analysis of Maximum Likelihood Parameter Estimates

                        Standard   95% Confidence     Chi-
   Parameter     DF Estimate  Error      Limits     Square Pr > ChiSq

   Intercept      1   3.9906  0.4374   3.1333   4.8479  83.23   <.0001
   fin            1   0.2837  0.1440   0.0015   0.5659   3.88   0.0488
   age            1   0.0425  0.0167   0.0098   0.0753   6.48   0.0109
   race           1  -0.2343  0.2298  -0.6846   0.2161   1.04   0.3079
   wexp           1   0.1106  0.1582  -0.1994   0.4206   0.49   0.4845
   mar            1   0.3246  0.2853  -0.2345   0.8837   1.29   0.2552
   paro           1   0.0618  0.1457  -0.2238   0.3474   0.18   0.6714
   prio           1  -0.0685  0.0218  -0.1113  -0.0257   9.83   0.0017
   Scale          1   0.7435  0.0665   0.6239   0.8861
   Weibull Shape  1   1.3449  0.1204   1.1285   1.6028
```

GENERATING PREDICTIONS AND HAZARD FUNCTIONS

After fitting a model with PROC LIFEREG, it's sometimes desirable to generate predicted survival times for the observations in the data set. If you want a single point estimate for each individual, the predicted median survival time is probably the best. You can get this easily with the OUTPUT statement, as shown in the following example:

```
PROC LIFEREG DATA=recid;
   MODEL week*arrest(0)=fin age race wexp mar paro prio
         / D=WEIBULL;
   OUTPUT OUT=a P=median STD=s;
RUN;
```

```
PROC PRINT DATA=a;
   VAR week arrest _prob_ median s;
RUN;
```

The P= option in the OUTPUT statement requests percentiles. By default, PROC LIFEREG calculates the 50th percentile (that is, the median). (The word *median* in the OUTPUT statement is just the variable name I chose to hold the quantiles.) You can request other percentiles with the QUANTILE keyword, as described in the PROC LIFEREG documentation. The STD keyword requests the standard errors of the medians.

Output 4.13 shows the first 20 cases in the new data set. In the output, _PROB_ is the quantile (that is, the percentile divided by 100) requested, and S is the standard error of the median. Note that many of these predicted medians are much greater than the observed event (or censoring times) and often well beyond the observation limit of 52 weeks. This is not surprising given that nearly 75 percent of the observations are censored. Although we observed individuals for only 52 weeks, the fitted distribution is not at all limited by the censoring times. Therefore, you should be very cautious in interpreting these predicted medians because the model is being extrapolated to times that are far beyond those that are actually observed.

Output 4.13 *Predicted Median Survival Times for Recidivism Data (First 20 Cases)*

Obs	week	arrest	_PROB_	median	s
1	20	1	0.5	86.910	14.532
2	17	1	0.5	43.353	6.646
3	25	1	0.5	45.257	14.654
4	52	0	0.5	167.947	50.293
5	52	0	0.5	87.403	21.731
6	52	0	0.5	86.166	13.989
7	23	1	0.5	148.225	42.416
8	52	0	0.5	93.083	15.532
9	52	0	0.5	54.873	7.694
10	52	0	0.5	83.518	15.483
11	52	0	0.5	109.544	18.986
12	52	0	0.5	165.290	50.312
13	37	1	0.5	56.411	10.132
14	52	0	0.5	146.285	38.768
15	25	1	0.5	65.358	8.360
16	46	1	0.5	110.592	18.115
17	28	1	0.5	45.471	6.814
18	52	0	0.5	65.816	10.306
19	52	0	0.5	44.305	9.429
20	52	0	0.5	122.208	30.552

You can also get predicted values for sets of covariate values that are not in the original data set. Before estimating the model, simply append to the data set additional observations with the desired covariate values and with the event time set to missing. These added observations are not used in estimating the model, but predicted values will be generated for them.

Instead of predicted survival times, researchers often want to predict the probability of surviving to some specified time (for example, 5-year survival probabilities). While these are not directly computed by PROC LIFEREG, it's fairly straightforward to calculate them. This is accomplished by substituting linear predictor values that can be produced by the OUTPUT statement into formulas for the survivor function (given in the PROC LIFEREG documentation). To make it easy, I've written a macro called PREDICT, which is described in detail in Appendix 1, "Macro Programs." This macro is used in the following way. When specifying the model in PROC LIFEREG, you must request that the parameter estimates be written to a data set using the OUTEST= option. Next, use the XBETA= option in the OUTPUT statement to request that the linear predictor be included in a second data set. Finally, call the macro, indicating the names of the two data sets, the name assigned to the linear predictor, and the time for calculating the survival probabilities. For example, to produce 30-week survival probabilities for the recidivism data, submit these statements:

```
PROC LIFEREG DATA=recid OUTEST=a;
   MODEL week*arrest(0) = fin age race wexp mar paro prio
         / D=WEIBULL;
   OUTPUT OUT=b XBETA=lp;
RUN;

%PREDICT(OUTEST=a,OUT=b,XBETA=lp,TIME=30)
```

Output 4.14 shows the first 20 cases of the new data set (_PRED_). The last column (prob) contains the 30-week survival probabilities based on the fitted model.

Output 4.14 *Predicted 30-Week Survival Probabilities for Recidivism Data*

Obs	week	fin	age	race	wexp	mar	paro	prio	educ	arrest	t	prob
1	20	0	27	1	0	0	1	3	3	1	30	0.85579
2	17	0	18	1	0	0	1	8	4	1	30	0.66139
3	25	0	19	0	1	0	1	13	3	1	30	0.67760
4	52	1	23	1	1	1	1	1	5	0	30	0.94010
5	52	0	19	0	1	0	1	3	3	0	30	0.85684
6	52	0	24	1	1	0	0	2	4	0	30	0.85417
7	23	0	25	1	1	1	1	0	4	1	30	0.92903
8	52	1	21	1	1	0	1	4	3	0	30	0.86811
9	52	0	22	1	0	0	0	6	3	0	30	0.74306
10	52	0	20	1	1	0	0	0	5	0	30	0.84816
11	52	1	26	1	0	0	1	3	3	0	30	0.89356
12	52	0	40	1	1	0	0	2	5	0	30	0.93878
13	37	0	17	1	1	0	1	5	3	1	30	0.75151
14	52	0	37	1	1	0	0	2	3	0	30	0.92776
15	25	0	20	1	0	0	1	3	4	1	30	0.79267
16	46	1	22	1	1	0	1	2	3	1	30	0.89490
17	28	0	19	1	0	0	0	7	3	1	30	0.67934
18	52	0	20	1	0	0	0	2	3	0	30	0.79448
19	52	0	25	1	0	0	1	12	3	0	30	0.66965
20	52	0	24	0	1	0	1	1	3	0	30	0.90800

Because every model estimated in PROC LIFEREG has an implicit hazard function, it would be nice to see what that hazard function looks like. I've written another macro called LIFEHAZ (also described in Appendix 1) that produces a graph of the hazard as a function of time. As with the PREDICT macro, you first need to fit a PROC LIFEREG model that includes the OUTPUT statement and the OUTEST= option in the PROC statement. Using the same PROC LIFEREG specification that was used before with the PREDICT macro, you then submit the following:

```
%LIFEHAZ(OUTEST=a,OUT=b,XBETA=lp)
```

For the recidivism data using the Weibull model, this macro produces the graph shown in Output 4.15. Keep in mind that this graph depends heavily on the specified model. If we specify a log-normal model before using the LIFEHAZ macro, the graph will look quite different.

Output 4.15 *Graph of Hazard Function for Recidivism Data*

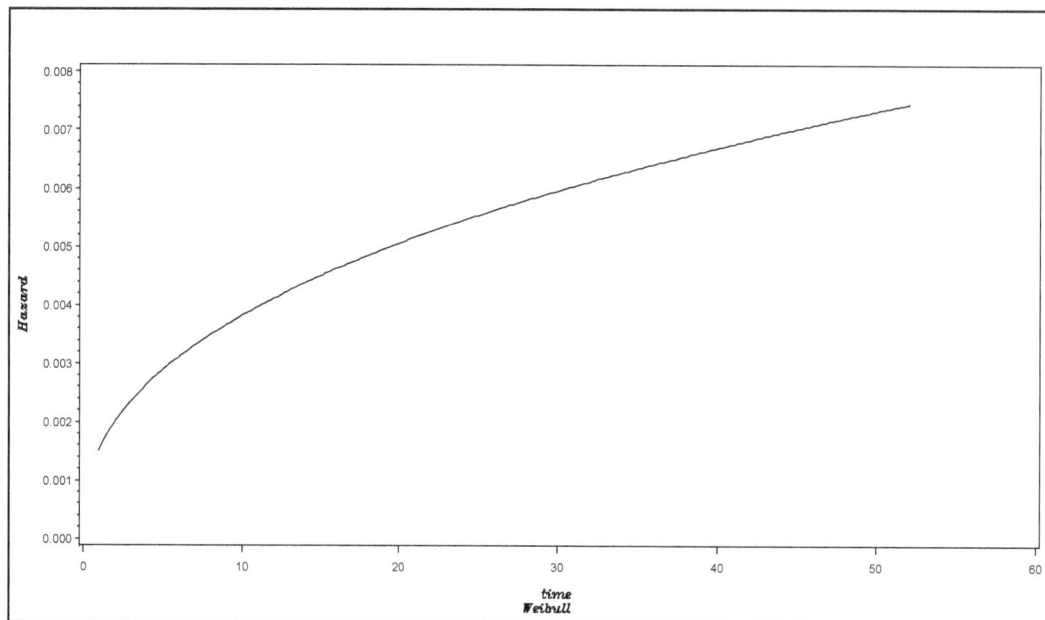

THE PIECEWISE EXPONENTIAL MODEL

All the AFT models we have considered so far assume that the hazard is a smooth, relatively simple function of time. The Cox model (estimated with the PHREG procedure) is much less restrictive in this regard, but it lacks the facility to test hypotheses about the shape of the hazard function. One way to get some of the flexibility of the Cox model without losing the hypothesis testing capability is to employ the piecewise exponential model, a method that is widely used in several fields. We can easily estimate it with PROC LIFEREG, although it requires some preliminary restructuring of the data. A bonus of this method is the ability to incorporate time-dependent covariates.

The basic idea is simple. Divide the time scale into intervals. Assume that the hazard is constant within each interval but can vary across intervals. In symbols, we define a set of J intervals, with cut points $a_0, a_1, \ldots, a_J$, where $a_0 = 0$, and $a_J = \infty$. Thus, interval j is given by $[a_{j-1}, a_j)$. The hazard for individual i is assumed to have the form

$$h_i(t) = \lambda_j e^{\beta \mathbf{x}_i} \quad \text{for } a_{j-1} \leq t < a_j$$

or equivalently

$$\log h_i(t) = \alpha_j + \beta \mathbf{x}_i$$

where $\alpha_j = \log \lambda_j$. Thus, the intercept in the log-hazard equation is allowed to vary in an unrestricted fashion from one interval to another. The choice of intervals is arbitrary, however, leading to some arbitrariness in the estimates.

The procedure for estimating this model is best explained by way of example. For the recidivism data, let's break up the 52-week observation period into four quarters of 13 weeks each and assume that the hazard is constant within each quarter. We then create a new data set with possibly multiple records for each person. One record is created for each quarter during which an individual was at risk of the first arrest. Four records are created for persons who were arrested in the fourth quarter or who were not arrested at all. Three records are created for those arrested in the third quarter, two records are created for those arrested in the second quarter, and one record is created for those arrested in the first quarter. This yields a total of 1,573 person-quarters for the 432 ex-convicts.

Each record is treated as a distinct observation, with the time reset to 0 at the beginning of the quarter. If an arrest occurred in the quarter, a censoring variable for that person-quarter is set to 1; otherwise, it is set to 0. If no arrest occurred in the quarter, a time variable is assigned the full 13 weeks. If an arrest occurred, the time variable is coded as the length of time from the start of the quarter until the arrest.

For the recidivism data, all censoring is at the end of the fourth quarter. Had there been any censoring within quarters, the time variable would be coded as the length of time from the beginning of the quarter until censoring occurred. The fixed covariates are simply replicated for each quarter. If there were any time-dependent covariates, their values at the beginning of each quarter could be assigned to the records for that quarter.

Here's a DATA step for creating such a data set:

```
DATA quarter;
   SET recid;
   quarter=CEIL(week/13);
   DO j=1 TO quarter;
      time=13;
      event=0;
       IF j=quarter AND arrest=1 THEN DO;
         event=1;
         time=week-13*(quarter-1);
      END;
      OUTPUT;
   END;
RUN;
```

The CEIL function, which produces the smallest integer greater than its argument, yields values of 1, 2, 3, or 4, corresponding to the quarter. The DO loop produces a record for each quarter at risk. The TIME and EVENT variables are initialized at the values appropriate for quarters in which arrests did not occur. The IF statement checks to see if an arrest occurred in the quarter. If an arrest did occur, then EVENT and TIME are appropriately recoded. Finally, a new record is output containing all the original variables plus the newly created ones.

We then call the LIFEREG procedure and specify an exponential model:

```
PROC LIFEREG DATA=quarter;
   CLASS j;
   MODEL time*event(0)=fin age race wexp mar paro prio j
        / D=EXPONENTIAL COVB;
RUN;
```

The variable J—the index variable in the DO loop of the DATA step—has values of 1, 2, 3, or 4, corresponding to the quarter covered by each record. It is specified as a CLASS variable so that PROC LIFEREG will set up an appropriate set of indicator variables to estimate the α_j's in the piecewise exponential model (actually we estimate contrasts between the α_j's).

Results in Output 4.16 show a significant effect of J (quarter), implying that the hazard is not constant over time. The Wald chi-square value is 8.70 on 3 d.f., which is corroborated by a likelihood-ratio test with a chi-square value of 9.52. (The likelihood-ratio test is calculated by rerunning the model without J and taking twice the positive difference in the log-likelihoods.) The coefficients for the three indicator variables are all contrasts with the fourth quarter. To interpret these coefficients, it's probably best to change their signs so that they reflect hazards rather than survival times. In contrast to the Weibull model, which imposed a monotonically increasing hazard, the pattern displayed here is not monotonic. The estimated hazard increases from first to second quarter, then decreases, and then increases again.

Output 4.16 *Results for Piecewise Exponential Model Applied to Recidivism Data*

```
                   Type III Analysis of Effects

                                   Wald
             Effect        DF    Chi-Square    Pr > ChiSq

               j            3      8.6954         0.0336
```

```
             Analysis of Maximum Likelihood Parameter Estimates

                            Standard   95% Confidence   Chi-
Parameter        DF Estimate  Error       Limits      Square Pr > ChiSq

Intercept        1   3.7226  0.6082   2.5307  4.9146   37.47   <.0001
fin              1   0.3774  0.1913   0.0023  0.7524    3.89   0.0486
age              1   0.0570  0.0220   0.0140  0.1000    6.73   0.0095
race             1  -0.3126  0.3080  -0.9162  0.2910    1.03   0.3101
wexp             1   0.1489  0.2122  -0.2670  0.5648    0.49   0.4828
mar              1   0.4331  0.3818  -0.3152  1.1814    1.29   0.2566
paro             1   0.0836  0.1957  -0.3000  0.4672    0.18   0.6692
prio             1  -0.0909  0.0286  -0.1470 -0.0347   10.07   0.0015
j          1     1   0.8202  0.2841   0.2633  1.3771    8.33   0.0039
j          2     1   0.1883  0.2446  -0.2911  0.6677    0.59   0.4414
j          3     1   0.3134  0.2596  -0.1953  0.8221    1.46   0.2273
j          4     0   0.0000    .         .       .       .       .
Scale            0   1.0000  0.0000   1.0000  1.0000
```

The chi-square tests for the individual indicator variables show that the hazard of arrest in the first quarter is significantly lower than the hazard in the last quarter. Although the two middle quarters have lower estimated hazards than the last, the differences are not significant. A Wald test (constructed from the covariance matrix produced by the COVB option) comparing the first and second quarters is also significant at about the .03 level. The coefficients and *p*-values for the remaining variables are consistent with those found with the conventional exponential and Weibull models.

Of course, there is a certain arbitrariness that arises from the division of the observation period into quarters. To increase confidence in the results, you may want to try different divisions and see if the results are stable. I re-estimated the model for the recidivism data with a division into 13 "months" of four weeks each, simply by changing all the 13's to 4's in the DATA step. This produces a data set with 4,991 records. Results for the fixed covariates are virtually identical. The Wald chi-square test for the overall effect of J (month) is also about the same, but, with 12 degrees of freedom, the *p*-value is well above conventional levels for statistical significance.

Here are some final observations about the piecewise exponential model:

- You do not need to be concerned about the fact that the working data set has multiple records for each individual. In particular, there is no inflation of test statistics resulting from lack of independence. The fact that the results are so similar regardless of how many observations are created should reassure you on this issue. The reason it's not a problem is that the likelihood function actually factors into a distinct term for each individual interval. This conclusion does *not* apply, however, when the data set includes multiple *events* for each individual.

- The piecewise model can be easily fit in a Bayesian framework using PROC PHREG. This option will be discussed in the next chapter.

- The piecewise exponential model is very similar to the discrete-time methods described in Chapter 7, "Analysis of Tied or Discrete Data with PROC LOGISTIC." The principal difference is that estimation of the piecewise exponential model uses information on the exact timing of events, while the discrete-time methods are based on interval-censored data.

- There is no requirement that the intervals have equal length, although that simplifies the DATA step somewhat. Because there's some benefit in having roughly equal numbers of events occurring in each interval, this sometimes requires unequal interval lengths.

- The use of time-dependent covariates in the piecewise exponential model can substantially complicate the DATA step that creates the multiple records. For examples of how to do this, see Chapter 7 on discrete-time methods. There are also a number of issues about design and interpretation of studies with time-dependent covariates that are discussed in detail in Chapters 5 and 7.

BAYESIAN ESTIMATION AND TESTING

Beginning with SAS 9.2, PROC LIFEREG can do a Bayesian analysis for any of the models discussed in this chapter. In a Bayesian analysis, the parameters of the model are treated as random variables rather than as fixed characteristics of the population. A *prior* distribution must be specified for these parameters. This is a probability distribution that incorporates any previous knowledge or beliefs about the parameters. The goal of the analysis is to produce the *posterior* distribution of the parameters. The posterior distribution is obtained by using Bayes' theorem to combine the likelihood function of the data with the prior distribution. Once you have the posterior distribution for a particular parameter, the mean or median of that distribution can be used as a point estimate of the parameter, and appropriate percentiles of the distribution can be used as interval estimates. For a useful primer on Bayesian analysis in SAS, see the *SAS/STAT 9.2 User's Guide*, "Introduction to Bayesian Analysis Procedures."

A Bayesian approach has several attractions that are relevant here:

■ Inferences are exact, in the sense that they do not rely on large-sample approximations, and thus may be more accurate for small samples.

■ Information from prior research studies can be readily incorporated into the analysis.

■ Comparison of non-nested models can be accomplished in a systematic framework.

These advantages come with a major computational cost. The Markov Chain Monte Carlo (MCMC) methods used by PROC LIFEREG to simulate the posterior distribution are very intensive. For example, it took only about a tenth of a second on my laptop to estimate the log-normal model reported in Output 4.1. By contrast, a Bayesian analysis (with the default non-informative prior distributions) took more than *5 minutes*.

The other potential disadvantage of a Bayesian approach is the difficulty in coming up with suitable prior distributions for the parameters of the model. The default is to use non-informative priors that embody little or no information about the parameters. PROC LIFEREG uses uniform (constant) priors for the coefficients and a "just informative" gamma prior for the scale parameter. But if you use these defaults, results typically don't differ much from conventional maximum likelihood estimation. In fact, for the exponential model (which has the scale parameter fixed at 1), the maximum likelihood estimator for the coefficients is simply the mode of the posterior distribution. Although you have the option of choosing informative priors, this is not such an easy or straightforward task.

As with most software for Bayesian analysis, PROC LIFEREG uses the MCMC algorithm (more specifically, the Gibbs sampler) to simulate a large number of sequential random draws from the posterior distribution—by default, 12,000 draws. The first 2,000 (the "burn-in" iterations) are discarded in order to ensure that that the algorithm has converged to the correct posterior distribution. Even then, however, there is no guarantee that convergence has been attained. For that reason, PROC LIFEREG provides diagnostic plots and several test statistics to evaluate the convergence of the distribution.

If you are OK with the defaults, requesting a Bayesian analysis is very simple. Just include the BAYES statement. Here is how to do it for the Weibull model that we estimated earlier (Output 4.3), using the recidivism data:

```
ODS HTML;
ODS GRAPHICS ON;
PROC LIFEREG DATA=recid;
  MODEL week*arrest(0)=fin age race wexp mar paro prio /D=WEIBULL;
  BAYES;
RUN;
ODS GRAPHICS OFF;
ODS HTML CLOSE;
```

ODS GRAPHICS is used here so that PROC LIFEREG will produce a plot of the posterior distribution for each parameter, along with two diagnostic plots, a trace plot and an autocorrelation plot. Selected portions of the tabular results are shown in Output 4.17. Plots for one of the parameters (the coefficient of AGE) are shown in Output 4.18.

The first two tables in Output 4.17 describe the prior distributions for all the parameters. By default, the regression coefficients all have the "Constant" or uniform prior, and they are assumed to be independent of each other and the scale parameter. The scale parameter is assigned a gamma distribution with "hyperparameters" of .001 for both the scale and inverse-shape parameters. This is equivalent to using a gamma distribution with a mean of 1 and a very large standard deviation (approximately 32).

The "Initial Values of the Chain" table displays the starting values for the parameters. These are "posterior mode" estimates, which means that they are obtained by maximizing the likelihood function weighted by the prior distribution. The "Fit Statistics" table shows three fit measures that were described earlier in this chapter: AIC, AICC, and BICC. In fact, the values shown here are exactly the same as those reported for conventional maximum likelihood estimation.

What's new are the DIC and p_D statistics. DIC stands for deviance information criterion, a widely used measure of fit for Bayesian analysis. It's essentially a Bayesian analog to the AIC statistic, calculated as

$$DIC = -2\log L(\overline{\theta}) + 2p_D$$

where L is the likelihood function, $\overline{\theta}$ is the mean of the posterior distribution of the parameter vector θ, and p_D is the "effective number of parameters." It is calculated as

$$p_D = Mean[-2\log L(\theta)] + 2\log L(\overline{\theta})$$

where the mean is taken over the posterior distribution of θ. As with the AIC, the DIC is used primarily to compare models with different numbers of parameters.

The next table, "Posterior Summaries," is the heart of the analysis. For each parameter, we get the mean of the posterior distribution, based on the sample of 10,000 random draws from that distribution. Keep in mind that all the numbers in this table are simulation-based estimates, and therefore will be slightly different if we run the model again. But with 10,000 cases, the sampling error should be small. The next column is the standard deviation, which plays the same role as the standard error in conventional frequentist methodology. Finally, we get the 25th, 50th, and 75th percentiles of the posterior distribution. The 50th percentile is, of course, the median, and some may prefer this to the mean as a point estimate for the parameter of interest.

The "Posterior Intervals" table displays what are usually called credible intervals in Bayesian analysis, and by default, they are 95% intervals. Credible intervals play the same role as confidence intervals in conventional statistics, although the interpretation is somewhat different. A 95% confidence interval is interpreted by saying that if we repeated the sampling and estimation process many times, 95% of the constructed confidence intervals would include the true parameter. A 95% credible interval says simply that there is a .95 probability that the true parameter is included in the reported interval.

The "Equal Tail Intervals" are merely the 2.5th percentiles and the 97.5th percentiles of the posterior distribution. Alternatively, one can use the "highest posterior density" or HPD intervals. These are the narrowest possible intervals that contain, by default, 95% of the posterior distribution. Although *p*-values are not reported for the estimates, a 95% credible interval that does not include 0 is the Bayesian equivalent of an estimate that is significantly different from 0 at the .05 level (by a two-sided test).

The two remaining tables are used as diagnostics for the performance of the MCMC algorithm. In the sequence of random draws from the posterior distribution, each new draw depends, in part, on the value of the preceding draw. That induces an autocorrelation in the drawn values that, ideally, should decline to 0 as the distance between draws gets larger. Persistently high values of the autocorrelation suggest that the algorithm is not mixing well. These autocorrelations can vary substantially from one parameter to another. For example, the autocorrelation for adjacent values (lag 1) of the "fin" coefficient is only .04, while for "age" it is .79. The "age" autocorrelation declines to .29 for values that are five draws apart (lag 5) and then to .08 for lag 10. A graph of the autocorrelations for the "age" coefficient is shown in Output 4.18.

Although autocorrelation should not cause any problem for estimates of the means, medians, and credible intervals, it may induce some bias in estimates of the standard deviations of the posterior distribution (Daniels and Hogan, 2008). This can be fixed by thinning the sample (for example, retaining only every 10th sample instead of every sample). To do this in PROC LIFEREG, you simply use THIN=10 as an option in the BAYES statement. Of course, the disadvantage of this is that you would need to generate 10 times as many samples to get the same degree of accuracy. And in my (limited) experience, using this option typically has very little impact on the standard deviations. If you do choose to thin, the autocorrelation graph should provide useful guidance on what level of thinning is needed.

The last table displays the "Geweke Diagnostics" for convergence of MCMC. The idea behind these statistics is very simple. After discarding the 2,000 burn-in iterations, we divide the sample into equal subsamples of early and late iterations. For the two subsamples, we calculate the means for each parameter. If the distribution has converged, these means should be equal, apart from sampling error. The Geweke statistic is just a standard z-statistic for differences between the two means. For the recidivism data, none of the differences approaches statistical significance. PROC LIFEREG also offers several other optional tests of convergence.

Output 4.17 *Results for Bayesian Analysis of the Weibull Model for Recidivism*

```
                          Bayesian Analysis

              Uniform Prior for Regression Coefficients

                       Parameter    Prior


                       Intercept    Constant
                       fin          Constant
                       age          Constant
                       race         Constant
                       wexp         Constant
                       mar          Constant
                       paro         Constant
                       prio         Constant



          Independent Prior Distributions for Model Parameters


                   Prior
   Parameter       Distribution                Hyperparameters

   Scale           Gamma          Shape     0.001    Inverse Scale     0.001



                      Initial Values of the Chain

      Chain       Seed  Intercept        fin        age       race       wexp

         1   681742001   3.989018    0.27029   0.040422   -0.22319   0.105745

                      Initial Values of the Chain

                 mar        paro         prio       Scale

            0.308981    0.058355     -0.06541    0.706806

                          Fit Statistics

          AIC (smaller is better)                  656.753
          AICC (smaller is better)                 657.180
          BIC (smaller is better)                  693.369
          DIC (smaller is better)                  657.110
          pD (effective number of parameters)        8.753
```

(*continued*)

Output 4.17 *(continued)*

Posterior Summaries

Parameter	N	Mean	Standard Deviation	Percentiles 25%	Percentiles 50%	Percentiles 75%
Intercept	10000	4.0183	0.4505	3.7214	4.0233	4.3152
fin	10000	0.2927	0.1502	0.1905	0.2909	0.3902
age	10000	0.0444	0.0171	0.0325	0.0438	0.0554
race	10000	-0.2661	0.2408	-0.4222	-0.2565	-0.1008
wexp	10000	0.1171	0.1648	0.00599	0.1161	0.2246
mar	10000	0.3767	0.3060	0.1647	0.3613	0.5749
paro	10000	0.0628	0.1523	-0.0384	0.0624	0.1625
prio	10000	-0.0675	0.0228	-0.0823	-0.0672	-0.0527
Scale	10000	0.7663	0.0708	0.7166	0.7619	0.8109

Posterior Intervals

Parameter	Alpha	Equal-Tail Interval		HPD Interval	
Intercept	0.050	3.1265	4.8962	3.1210	4.8881
fin	0.050	0.00557	0.5971	0.000220	0.5888
age	0.050	0.0130	0.0802	0.0126	0.0796
race	0.050	-0.7700	0.1813	-0.7401	0.2044
wexp	0.050	-0.2005	0.4436	-0.1940	0.4487
mar	0.050	-0.1900	0.9996	-0.1935	0.9931
paro	0.050	-0.2351	0.3628	-0.2255	0.3686
prio	0.050	-0.1137	-0.0225	-0.1139	-0.0229
Scale	0.050	0.6383	0.9201	0.6295	0.9067

Posterior Autocorrelations

Parameter	Lag 1	Lag 5	Lag 10	Lag 50
Intercept	0.5590	0.0933	-0.0033	-0.0137
fin	0.0437	0.0064	0.0134	-0.0067
age	0.7862	0.2927	0.0821	-0.0099
race	0.6416	0.1024	0.0067	0.0076
wexp	0.2374	0.0563	0.0047	-0.0096
mar	0.7769	0.2421	0.0207	-0.0073
paro	0.0808	0.0139	0.0183	-0.0040
prio	0.7405	0.1581	0.0114	-0.0093
Scale	0.6061	0.1295	0.0677	-0.0145

(continued)

Output 4.17 (continued)

```
                    Geweke Diagnostics

        Parameter          z     Pr > |z|

        Intercept       0.8086     0.4188
        fin             1.4737     0.1406
        age            -0.0764     0.9391
        race           -0.4365     0.6625
        wexp           -0.6223     0.5337
        mar             0.7653     0.4441
        paro           -0.0576     0.9541
        prio           -1.0156     0.3098
        Scale           1.3244     0.1854
```

Output 4.18 displays the plots that are produced by default when the BAYES statement is used. Only the plots for the AGE coefficient are shown. We have already considered the autocorrelation plot. The top plot is a trace plot of the sampled values of the AGE coefficient across the sequence of iterations, from 2,000 to 12,000. The first 2,000 values are not shown because they are considered burn-in iterations and are not used in calculating any of the estimates. Ideally, the trace plot should look like white noise (that is, it should have no pattern whatsoever). Certainly, we don't want to see any upward or downward trends, nor any evidence that the variability is increasing or decreasing. The plot shown here is consistent with the relatively high levels of autocorrelation that persist up to about lag 10.

The last plot is a graph of the empirical density function for the posterior distribution. These graphs typically look similar to normal distributions, especially if the sample size is large. In this example, the bulk of the distribution is clearly well above 0, consistent with the 95% credible intervals in Output 4.17 that do not include zero.

Output 4.18 *Graphs for Bayesian Analysis of Recidivism Data*

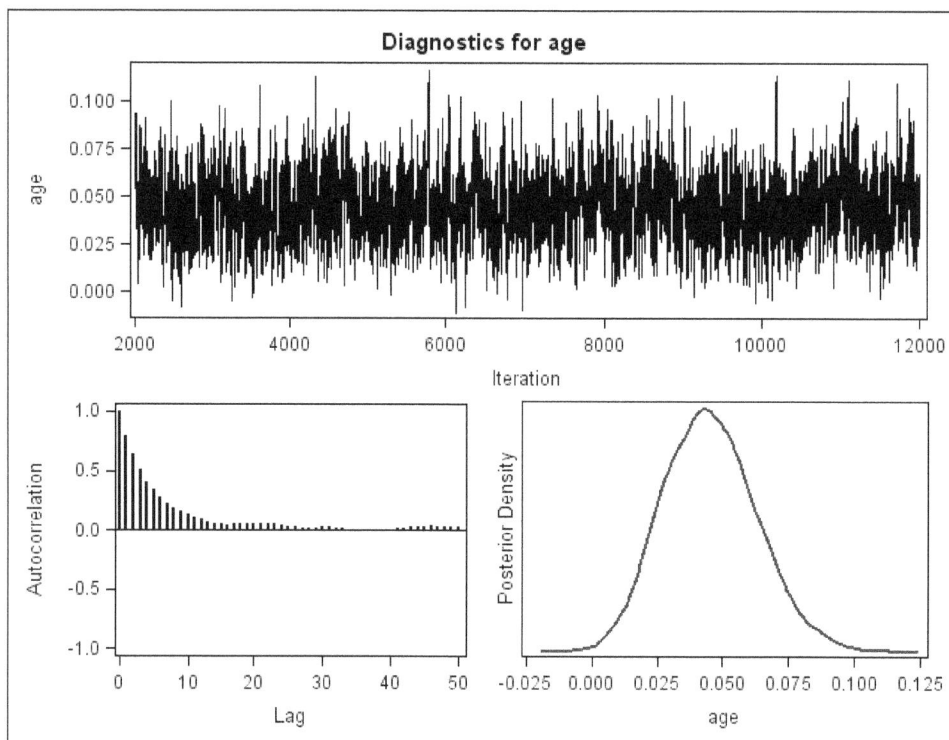

CONCLUSION

PROC LIFEREG provides effective methods for regression analysis of censored survival data, especially data with left censoring or interval censoring. These methods are somewhat less robust than the more widely used Cox regression analysis performed by PROC PHREG but, in most cases, the results produced by the two approaches are very similar. Moreover, unlike PROC PHREG, PROC LIFEREG makes it possible to test certain hypotheses about the shape of the hazard function.

The biggest limitation of the PROC LIFEREG models is the inability to incorporate time-dependent covariates, although you can accomplish this to some degree with the piecewise exponential model. We now turn to PROC PHREG, which excels at this particular task.

CHAPTER **5**
Estimating Cox Regression Models with PROC PHREG

INTRODUCTION

PROC PHREG (pronounced P-H-REG, not FREG) implements the regression method first proposed in 1972 by the British statistician Sir David Cox in his famous paper "Regression Models and Life Tables" (*Journal of the Royal Statistical Society, Series B*). It's difficult to exaggerate the impact of this paper. In the 2009 *ISI Web of Science*, it was cited over 1,000 times, making it the most highly cited journal article in the entire literature of statistics. In fact, Garfield (1990) reported that its cumulative citation count placed it among the top 100 papers in all of science. These citation counts undoubtedly underestimate the actual use of the method because many authors don't bother citing the original paper.

What explains this enormous popularity? Perhaps the most important reason is that, unlike the parametric methods discussed in Chapter 4, "Estimating Parametric Regression Models with PROC LIFEREG," Cox's method does not require that you choose some particular probability distribution to represent survival times. That's why it's called *semi*parametric. As a consequence, Cox's method (often referred to as *Cox* *regression*) is considerably more robust. A second reason for the paper's popularity is that Cox regression makes it relatively easy to incorporate time-dependent covariates (that is, covariates that may change in value over the course of the observation period).

There are other attractive features of Cox regression that are less widely known or appreciated. Cox regression permits a kind of stratified analysis that is very effective in controlling for nuisance variables. And Cox regression makes it easy to adjust for periods of time in which an individual is not at risk of an event. Finally, Cox regression can readily accommodate both discrete and continuous measurement of event times.

Despite all these desirable qualities, Cox regression should not be viewed as the universal method for regression analysis of survival data. As I indicated in Chapter 4, there are times when a parametric method is preferable. And for most applications, you can do a reasonably good job of survival analysis using only PROC LIFEREG and PROC LIFETEST. Still, if I could have only one SAS procedure for doing survival analysis, it would be PROC PHREG.

All implementations of Cox regression are not created equal. Among the many available commercial programs, PROC PHREG stands out for its toolkit of powerful features. While some programs don't allow stratification—a fatal deficit in my view—PROC PHREG has a very flexible stratification option. Many programs don't handle time-dependent covariates at all; those that do often have severe restrictions on the number or kinds of such covariates. In contrast, PROC PHREG has by far the most extensive and powerful capabilities for incorporating time-dependent covariates. And while Cox regression can theoretically deal with discrete (*tied*) data, most programs use approximations that are inadequate in many cases. In contrast, PROC PHREG offers two exact algorithms for tied data.

THE PROPORTIONAL HAZARDS MODEL

In his 1972 paper, Cox made two significant innovations. First, he proposed a model that is usually referred to as the *proportional hazards model*. That name is somewhat misleading, however, because the model can readily be generalized to allow for nonproportional hazards. Second, he proposed a new estimation method that was later named *partial likelihood* or, more accurately, *maximum partial likelihood*. The term *Cox regression* refers to the combination of the model and the estimation method. It didn't take any great leap of imagination to formulate the proportional hazards model—it's a relatively straightforward generalization of the Weibull and Gompertz models that we considered in Chapter 2, "Basic Concepts of Survival Analysis." But the partial likelihood method is something completely different. It took years for statisticians to fully understand and appreciate this novel approach to estimation.

Before discussing partial likelihood, let's first examine the model that it was designed to estimate. We'll start with the basic model that does not include time-dependent covariates or nonproportional hazards. The model is usually written as

$$h_i(t) = \lambda_0(t)\exp(\beta_1 x_{i1} + \ldots + \beta_k x_{ik}) \tag{5.1}$$

This equation says that the hazard for individual i at time t is the product of two factors:

- a function $\lambda_0(t)$ that is left unspecified, except that it can't be negative
- a linear function of a set of k fixed covariates, which is then exponentiated.

The function $\lambda_0(t)$ can be regarded as the hazard function for an individual whose covariates all have values of 0. It is often called the *baseline hazard function*.

Taking the logarithm of both sides, we can rewrite the model as

$$\log h_i(t) = \alpha(t) + \beta_1 x_{i1} + \ldots + \beta_k x_{ik} \tag{5.2}$$

where $\alpha(t) = \log \lambda_0(t)$. If we further specify $\alpha(t) = \alpha$, we get the exponential model. If we specify $\alpha(t) = \alpha t$, we get the Gompertz model. Finally, if we specify $\alpha(t) = \alpha \log t$, we have the Weibull model. As we will see, however, the great attraction of Cox regression is that such choices are unnecessary. The function $\alpha(t)$ can take any form whatever, even that of a step function.

Why is equation (5.1) called the proportional hazards model? Because the hazard for any individual is a fixed proportion of the hazard for any other individual. To see this, take the ratio of the hazards for two individuals i and j, and apply equation (5.1):

$$\frac{h_i(t)}{h_j(t)} = \exp\{\beta_1(x_{i1} - x_j) + \ldots + \beta_k(x_{ik} - x_{jk})\} \tag{5.3}$$

What's important about this equation is that $\lambda_0(t)$ cancels out of the numerator and denominator. As a result, the ratio of the hazards is constant over time. If we graph the log hazards for any two individuals, the proportional hazards property implies that the hazard functions should be strictly parallel, as in Figure 5.1.

Figure 5.1 *Parallel Log-Hazard Functions from Proportional Hazards Model*

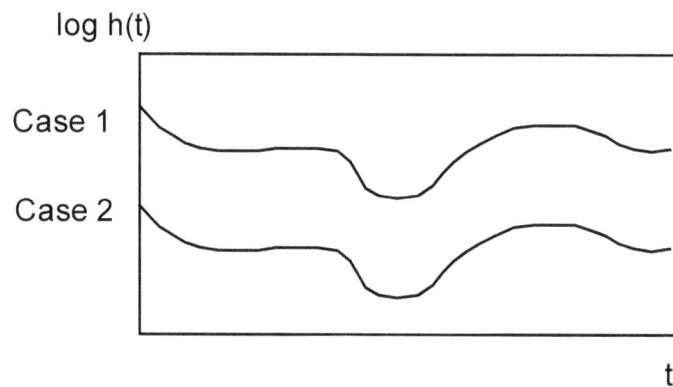

PARTIAL LIKELIHOOD

What's remarkable about partial likelihood is that you can estimate the β coefficients of the proportional hazards model without having to specify the baseline hazard function $\lambda_0(t)$. In this section, we will

- consider some general properties of partial likelihood
- look at two simple examples using PROC PHREG
- examine the mathematics of the method in some detail.

The likelihood function for the proportional hazards model of equation (5.1) can be factored into two parts:

- one part depends on both $\lambda_0(t)$ and $\beta = [\beta_1 \ \beta_2 \ ... \ \beta_k]'$, the vector of coefficients
- the other part depends on β alone.

What partial likelihood does, in effect, is discard the first part and treat the second part—the partial likelihood function—as though it were an ordinary likelihood function. You get estimates by finding values of β that maximize the partial likelihood. Because there is *some* information about β in the discarded portion of the likelihood function, the resulting estimates are not fully efficient. Their standard errors are larger than they would be if you used the entire likelihood function to obtain the estimates. In most cases, however, the loss of efficiency is quite small (Efron, 1977). What you gain in return is robustness because the estimates have good properties regardless of the actual shape of the baseline hazard function. To be specific, partial likelihood estimates still have two of the three standard properties of ML estimates: they are consistent and asymptotically normal. In other words, in large samples they are approximately unbiased and their sampling distribution is approximately normal.

Another interesting property of partial likelihood estimates is that they depend only on the *ranks* of the event times, not their numerical values. This implies that any monotonic transformation of the event times will leave the coefficient estimates unchanged. For example, we could add a constant to everyone's event time, multiply the result by a constant, take the logarithm, and then take the square root—all without producing the slightest change in the coefficients or their standard errors.

Partial Likelihood: Examples

Let's first apply the partial likelihood method to the recidivism data that we introduced in **The Life-Table Method** in Chapter 3, "Estimating and Comparing Survival Curves with PROC LIFETEST," and that was repeatedly analyzed in Chapter 4. The syntax for PROC PHREG is almost identical to that for PROC LIFEREG, except that you do not need to specify a distribution:

```
PROC PHREG DATA=recid;
   MODEL week*arrest(0)=fin age race wexp mar paro prio;
RUN;
```

Output 5.1 shows the results.

Output 5.1 *Results from Partial Likelihood Estimation with Recidivism Data*

```
                      The PHREG Procedure

                      Model Information

         Data Set                    MY.RECID
         Dependent Variable          week
         Censoring Variable          arrest
         Censoring Value(s)          0
         Ties Handling               BRESLOW


         Number of Observations Read          432
         Number of Observations Used          432


   Summary of the Number of Event and Censored Values

                                           Percent
         Total      Event     Censored     Censored

          432        114         318         73.61


                   Convergence Status

   Convergence criterion (GCONV=1E-8) satisfied.


                 Model Fit Statistics

                        Without            With
         Criterion     Covariates        Covariates

         -2 LOG L       1351.367          1318.241
         AIC            1351.367          1332.241
         SBC            1351.367          1351.395
```

(continued)

Output 5.1 *(continued)*

```
                Testing Global Null Hypothesis: BETA=0

         Test                Chi-Square      DF      Pr > ChiSq

         Likelihood Ratio      33.1256        7        <.0001
         Score                 33.3828        7        <.0001
         Wald                  31.9875        7        <.0001


                Analysis of Maximum Likelihood Estimates

                    Parameter    Standard                              Hazard
   Parameter   DF    Estimate      Error   Chi-Square  Pr > ChiSq      Ratio

   fin          1    -0.37902     0.19136    3.9228      0.0476        0.685
   age          1    -0.05724     0.02198    6.7798      0.0092        0.944
   race         1     0.31415     0.30802    1.0402      0.3078        1.369
   wexp         1    -0.15113     0.21212    0.5076      0.4762        0.860
   mar          1    -0.43280     0.38180    1.2850      0.2570        0.649
   paro         1    -0.08497     0.19575    0.1884      0.6642        0.919
   prio         1     0.09114     0.02863   10.1331      0.0015        1.095
```

The preliminary information is the same as in PROC LIFEREG, except for the line "Ties Handling: BRESLOW." This line refers to the default method for handling *ties*—two or more observations that have exactly the same event time. Although Breslow's method is nearly universal, we'll consider three superior alternatives later in the **Tied Data** section. The next table, labeled "Model Fit Statistics," displays measures of fit that were described in Chapter 4:

- -2 times the log likelihood
- Akaike's information criterion
- Schwarz's Bayesian criterion (equivalent to the Bayesian information criterion [BIC] statistic reported by PROC LIFEREG)

The difference here is that these statistics are based on the partial likelihood rather than the full likelihood. Notice that the three fit measures are the same for a model with no covariates. This is not surprising because the main purpose of Akaike's information criterion (AIC) and Schwartz's Bayesian criterion (SBC) statistics is to penalize models for having more covariates, allowing one to compare non-nested models with different numbers of covariates.

The table labeled "Testing Global Null Hypothesis: BETA=0" displays tests of the null hypothesis that all the coefficients are 0. Three alternative chi-square statistics are given: a likelihood-ratio test, a score test, and a Wald test. The likelihood ratio chi-square is just the difference in the −2 log L values for the models with and without covariates shown in the previous table. I already discussed the general properties of these tests in Chapter 4 (see **Hypothesis Tests**). Here we see that all three statistics are a bit over 30 with 7 d.f. leading to very small *p*-values. (The 7 d.f. correspond to the seven coefficients in the model.) We conclude that at least one of the coefficients is not 0.

In the last table, we see the coefficient estimates and associated statistics. Notice that there is no intercept estimate—a characteristic feature of partial likelihood estimation. The intercept is part of $\alpha(t)$, the arbitrary function of time, which drops out of the estimating equations. As with PROC LIFEREG, the chi-square tests are Wald tests for the null hypothesis that each coefficient is equal to 0. These statistics are calculated simply by squaring the ratio of each coefficient to its estimated standard error. The last column, labeled "Hazard Ratio," is just exp(β).

Hazard ratios can be interpreted almost exactly like odds ratios in logistic regression. For indicator (dummy) variables with values of 1 and 0, you can interpret the hazard ratio as the ratio of the estimated hazard for those with a value of 1 to the estimated hazard for those with a value of 0 (controlling for other covariates). For example, the estimated hazard ratio for the variable FIN (financial aid) is .685. This means that the hazard of arrest for those who received financial aid is only about 69 percent of the hazard for those who did not receive aid (controlling for other covariates).

For quantitative covariates, a more helpful statistic is obtained by subtracting 1.0 from the hazard ratio and multiplying by 100. This gives the estimated percent change in the hazard for each 1-unit increase in the covariate. For the variable AGE, the hazard ratio is .944, which yields $100(.944 − 1) = − 5.6$. Therefore, for each 1-year increase in the age at release, the hazard of arrest goes down by an estimated 5.6 percent.

Overall, the results are similar to those we saw in Chapter 4 with the LIFEREG procedure. There are highly significant effects of age and the number of prior convictions and a marginally significant effect of financial aid. Comparing the coefficients with those in Output 4.2 for the exponential model, we find that all the numbers are very close, but the signs are reversed. The *p*-values are also similar. The sign reversal is not surprising because the PROC LIFEREG estimates are in log-survival time format, while the PROC PHREG estimates are in log-hazard format. The PROC PHREG estimates are all larger in magnitude than the Weibull estimates in Output 4.3, but, again, that's merely a consequence of the alternative ways of expressing the model. When we convert the Weibull estimates to log-hazard format by dividing by the scale estimate and changing the sign (as in **The Weibull Model** section in Chapter 4), the results are remarkably close to the PROC PHREG estimates. Because the other PROC LIFEREG models (gamma, log-logistic, and log-normal) are not proportional hazards models, their coefficients cannot be converted to log-hazard format. Consequently, there is no point in comparing them with the PROC PHREG coefficients.

Now let's look at a somewhat more complicated example, the famous Stanford Heart Transplant Data, as reported by Crowley and Hu (1977). The sample consists of 103 cardiac patients who were enrolled in the transplantation program between 1967 and 1974. After enrollment, patients waited varying lengths of time until a suitable donor heart was found. Thirty patients died before receiving a transplant, while another four patients had still not received transplants at the termination date of April 1, 1974. Patients were followed until death or until the termination date. Of the 69 transplant recipients, only 24 were still alive at termination. At the time of transplantation, all but four of the patients were tissue typed to determine the degree of similarity with the donor.

The following variables were input into SAS:

DOB   date of birth.
DOA   date of acceptance into the program.
DOT   date of transplant.
DLS   date last seen (dead or censored).
DEAD   coded 1 if dead at DLS; otherwise, coded 0.
SURG   coded 1 if patient had open-heart surgery prior to DOA; otherwise, coded 0.
M1   number of donor alleles with no match in recipient (1 through 4).
M2   coded 1 if donor-recipient mismatch on HLA-A2 antigen; otherwise, 0.
M3   mismatch score.

The variables DOT, M1, M2, and M3 are coded as missing for those patients who did not receive a transplant. All four date measures are coded in the form *mm/dd/yy*, where *mm* is the month, *dd* is the day, and *yy* is the year. Here are the raw data for the first 10 cases:

DOB	DOA	DOT	DLS	DEAD	SURG	M1	M2	M3
01/10/37	11/15/67	.	01/03/68	1	0	.	.	.
03/02/16	01/02/68	.	01/07/68	1	0	.	.	.
09/19/13	01/06/68	01/06/68	01/21/68	1	0	2	0	1.110
12/23/27	03/28/68	05/02/68	05/05/68	1	0	3	0	1.660
07/28/47	05/10/68	.	05/27/68	1	0	.	.	.
11/08/13	06/13/68	.	06/15/68	1	0	.	.	.
08/29/17	07/12/68	08/31/68	05/17/70	1	0	4	0	1.320
03/27/23	08/01/68	.	09/09/68	1	0	.	.	.
06/11/21	08/09/68	.	11/01/68	1	0	.	.	.
02/09/26	08/11/68	08/22/68	10/07/68	1	0	2	0	0.610

These data were read into SAS with the following DATA step. (Note: The OPTIONS statement preceding the DATA step corrects a Y2K problem in the data.)

```
OPTIONS YEARCUTOFF=1900;
DATA stan;
   INFILE 'c: stan.dat';
   INPUT dob mmddyy9. doa mmddyy9. dot mmddyy9. dls mmddyy9.
         dead surg m1 m2 m3;
   surv1=dls-doa;
   surv2=dls-dot;
   ageaccpt=(doa-dob)/365.25;
   agetrans=(dot-dob)/365.25;
   wait=dot-doa;
   IF dot=. THEN trans=0; ELSE trans=1;
RUN;
```

Notice that the four date variables are read with the MMDDYY9. format, which translates the date into the number of days since January 1, 1960. (Dates earlier than that have negative values.) We then create two survival time variables: days from acceptance until death (SURV1) and days from transplant until death (SURV2). We also calculate the age (in years) at acceptance into the program (AGEACCPT), the age at transplant (AGETRANS), and the number of days from acceptance to transplant (WAIT). Finally, we create an indicator variable (TRANS) coded 1 for those who received a transplant and 0 for those who did not.

An obvious question is whether transplantation raised or lowered the hazard of death. A naive approach to answering this question is to do a Cox regression of SURV1 on transplant status (TRANS), controlling for AGEACCPT and SURG:

```
PROC PHREG DATA=stan;
   MODEL surv1*dead(0)=trans surg ageaccpt;
RUN;
```

The results in Output 5.2 show very strong effects of both transplant status and age at acceptance. We see that each additional year of age at the time of acceptance into the program is associated with a 6 percent increase in the hazard of death. On the other hand, the hazard for those who received a transplant is only about 18 percent of the hazard for those who did not (see the Hazard Ratio column). Or equivalently (taking the reciprocal), those who did not receive transplants are about 5-1/2 times more likely to die at any given point in time.

Output 5.2 *Results for All Patients, No Time-Dependent Variables*

```
             Testing Global Null Hypothesis: BETA=0

     Test                  Chi-Square      DF      Pr > ChiSq

     Likelihood Ratio        45.4629        3        <.0001
     Score                   52.0469        3        <.0001
     Wald                    46.6680        3        <.0001


             Analysis of Maximum Likelihood Estimates

                   Parameter   Standard                           Hazard
  Parameter   DF   Estimate      Error    Chi-Square   Pr > ChiSq   Ratio

  trans        1   -1.70813     0.27860    37.5902       <.0001     0.181
  surg         1   -0.42130     0.37098     1.2896       0.2561     0.656
  ageaccpt     1    0.05860     0.01505    15.1611       <.0001     1.060
```

While the age effect may be real, the transplant effect is almost surely an artifact. The main reason why patients did *not* get transplants is that they died before a suitable donor could be found. Thus, when we compare the death rates for those who did and did not get transplants, the rates are much higher for those who did not. In effect, the covariate is actually a *consequence* of the dependent variable: an early death prevents a patient from getting a transplant. The way around this problem is to treat transplant status as a time-dependent covariate, but that will have to wait until the **Time-Dependent Covariates** section later in this chapter.

We can also ask a different set of questions that do not require any time-dependent covariates. Restricting the analysis to the 65 patients who *did* receive heart transplants, we can ask why some of these patients survived longer than others:

```
PROC PHREG DATA=stan;
   WHERE trans=1;
   MODEL surv2*dead(0)=surg m1 m2 m3 agetrans wait dot;
RUN;
```

Notice that we now use a different origin—the date of the transplant—in calculating survival time. (It is possible to use the date of acceptance as the origin, using the methods in the **Left Truncation and Late Entry into the Risk Set** section later in this chapter, but it is probably not worth the trouble.)

Output 5.3 *Results for Transplant Patients, No Time-Dependent Covariates*

```
                  Testing Global Null Hypothesis: BETA=0

          Test                 Chi-Square     DF     Pr > ChiSq

          Likelihood Ratio       16.5855       7       0.0203
          Score                  15.9237       7       0.0258
          Wald                   14.9076       7       0.0372


                  Analysis of Maximum Likelihood Estimates

                      Parameter    Standard                         Hazard
   Parameter   DF     Estimate       Error   Chi-Square  Pr > ChiSq  Ratio

   surg         1     -0.77029     0.49718     2.4004     0.1213     0.463
   m1           1     -0.24857     0.19437     1.6355     0.2009     0.780
   m2           1      0.02958     0.44268     0.0045     0.9467     1.030
   m3           1      0.64407     0.34276     3.5309     0.0602     1.904
   agetrans     1      0.04927     0.02282     4.6619     0.0308     1.050
   wait         1     -0.00197     0.00514     0.1469     0.7015     0.998
   dot          1    -0.0001650   0.0002991    0.3044     0.5811     1.000
```

Results in Output 5.3 show, again, that older patients have higher risks of dying. Specifically, each additional year of age at the time of the transplant is associated with a 5 percent increase in the hazard of death. That does not tell us whether the *surgery* is riskier for older patients, however. It merely tells us that older patients are more likely to die. There is also some evidence of higher death rates for those who have a higher level of tissue mismatch, as measured by the M3 score. None of the other variables approaches statistical significance, however. (Note that four cases are lost in this analysis because of missing data on M1-M3.)

Partial Likelihood: Mathematical and Computational Details

Now that we've seen the partial likelihood method in action, let's take a closer look at how it does what it does. Using the same notation as in Chapter 4, we have n independent individuals ($i = 1,..., n$). For each individual i, the data consist of three parts: t_i, δ_i and $\mathbf{x}_i$, where t_i is the time of the event or the time of censoring, δ_i is an indicator variable with a value of 1 if t_i is uncensored or a value of 0 if t_i is censored, and $\mathbf{x}_i = [x_{i1} \ldots x_{ik}]$ is a vector of k covariate values.

An ordinary likelihood function is typically written as a product of the likelihoods for all the individuals in the sample. On the other hand, you can write the partial likelihood as a product of the likelihoods for all the *events* that are observed. Thus, if J is the number of events, we can write

$$PL = \prod_{j=1}^{J} L_j \tag{5.4}$$

where L_j is the likelihood for the jth event. Next we need to see how the individual L_js are constructed. This is best explained by way of an example. Consider the data in Output 5.4, which is taken from Collett (2003) with a slight modification (the survival time for observation 8 is changed from 26 to 25 to eliminate ties). The variable SURV contains the survival time in months, beginning with the month of surgery, for 45 breast cancer patients. Twenty-six of the women died (DEAD=1) during the observation period, so there are 26 terms in the partial likelihood. The variable X has a value of 1 if the tumor had a positive marker for possible metastasis; otherwise, the variable has a value of 0. The cases are arranged in ascending order by survival time, which is convenient for constructing the partial likelihood.

Output 5.4 *Survival Times for Breast Cancer Patients*

OBS	EVENT	SURV	DEAD	X
1	1	5	1	1
2	2	8	1	1
3	3	10	1	1
4	4	13	1	1
5	5	18	1	1
6	6	23	1	0
7	7	24	1	1
8	8	25	1	1
9	9	26	1	1
10	10	31	1	1

(*continued*)

Output 5.4 *(continued)*

11	11	35	1	1
12	12	40	1	1
13	13	41	1	1
14	14	47	1	0
15	15	48	1	1
16	16	50	1	1
17	17	59	1	1
18	18	61	1	1
19	19	68	1	1
20	20	69	1	0
21	.	70	0	0
22	21	71	1	1
23	.	71	0	0
24	.	76	0	1
25	.	100	0	0
26	.	101	0	0
27	.	105	0	1
28	.	107	0	1
29	.	109	0	1
30	22	113	1	1
31	.	116	0	1
32	23	118	1	1
33	24	143	1	1
34	25	148	1	0
35	.	154	0	1
36	.	162	0	1
37	26	181	1	0
38	.	188	0	1
39	.	198	0	0
40	.	208	0	0
41	.	212	0	0
42	.	212	0	1
43	.	217	0	1
44	.	224	0	0
45	.	225	0	1

The first death occurred to patient 1 in month 5. To construct the partial likelihood (L_1) for this event, we ask the following question: Given that a death occurred in month 5, what is the probability that it happened to patient 1 rather than to one of the other patients? The answer is the hazard for patient 1 at month 5 divided by the sum of the hazards for all the patients who were at risk of death in that same month. At month 5, all 45 patients were at risk of death, so the probability is

$$L_1 = \frac{h_1(5)}{h_1(5) + h_2(5) + \ldots + h_{45}(5)} \tag{5.5}$$

While this expression has considerable intuitive appeal, the derivation is actually rather involved and will not be presented here.

The second death occurred to patient 2 in month 8. Again we ask, given that a death occurred in month 8, what is the probability that it occurred to patient 2 rather than to one of the other patients at risk? Patient 1 is no longer at risk of death because she already died. So L_2 has the same form as L_1, but the hazard for patient 1 is removed from the denominator:

$$L_2 = \frac{h_2(8)}{h_2(8) + h_3(8) + \ldots + h_{45}(8)} \qquad (5.6)$$

The set of all individuals who are at risk at a given point in time is often referred to as the *risk set*. At time 8, the risk set consists of patients 2 through 45, inclusive.

We continue in this way for each successive death, deleting from the denominator the hazards for all those who have already died. Also deleted from the denominator are those who have been censored at an earlier point in time. That's because they are no longer at risk of an observed event. For example, the 21st death occurred to patient 22 in month 71. Patient 21 was censored at month 70, so her hazard does not appear in the denominator of L_{21}. On the other hand, if an event time is the same as a censoring time, the convention is to assume that the censored observation was still at risk at that time. Thus, patient 23 who was censored in month 71 *does* show up in the denominator of L_{21}.

The last term in the likelihood corresponds to the 26th death, which occurred to the 37th patient in month 181:

$$L_{26} = \frac{h_{37}(181)}{h_{37}(181) + h_{38}(181) + \ldots + h_{45}(181)} \qquad (5.7)$$

All the hazards in the denominator, except for the first, are for patients who were censored in months later than 181.

The results to this point have made no assumptions about the form of the hazard function. Now, we invoke the proportional hazards model of equation (5.1) and substitute the expression for the hazard into the expression for L_1,

$$L_1 = \frac{\lambda_0(5)e^{\beta x_1}}{\lambda_0(5)e^{\beta x_1} + \lambda_0(5)e^{\beta x_2} + \ldots + \lambda_0(5)e^{\beta x_{45}}} \qquad (5.8)$$

where x_i is the value of x for the ith patient. This leads to a considerable simplification because the unspecified function $\lambda_0(5)$ is common to every term in the expression. Canceling, we get

$$L_1 = \frac{e^{\beta x_1}}{e^{\beta x_1} + e^{\beta x_2} + \ldots + e^{\beta x_{45}}}. \qquad (5.9)$$

It is this cancellation of the λs that makes it possible to estimate the $\boldsymbol{\beta}$ coefficients without having to specify the baseline hazard function. Of course, the λs also cancel for all the other terms in the partial likelihood.

Earlier I remarked that the partial likelihood depends only on the order of the event times, not on their exact values. You can easily see this by considering each of the L_j terms. Although the first death occurred in month 5, L_1 would be exactly the same if it had occurred at any time from 0 up to (but not including) 8, the month of the second event. Similarly, L_2 would have been the same if the second death had occurred any time greater than 5 and less than 10 (the month of the third death).

A general expression for the partial likelihood for data with time-invariant covariates from a proportional hazards model is

$$PL = \prod_{i=1}^{n} \left(\frac{e^{\boldsymbol{\beta}\mathbf{x}_i}}{\sum_{j=1}^{n} Y_{ij} e^{\boldsymbol{\beta}\mathbf{x}_j}} \right)^{\delta_i} \tag{5.10}$$

where $Y_{ij} = 1$ if $t_j \geq t_i$; and $Y_{ij} = 0$ if $t_j < t_i$. (The Ys are just a convenient mechanism for excluding from the denominator those individuals who have already experienced the event and are, thus, not part of the risk set.) Although this expression has the product taken over all individuals rather than over all events (as in equation [5.4]), the terms corresponding to censored observations are effectively excluded because $\delta_i = 0$ for those cases. This expression is not valid for tied event times, but it does allow for ties between one event time and one or more censoring times.

Once the partial likelihood is constructed, you can maximize it with respect to $\boldsymbol{\beta}$ just like an ordinary likelihood function. As usual, it's convenient to maximize the logarithm of the likelihood, which is

$$\log PL = \sum_{i=1}^{n} \delta_i \left[\boldsymbol{\beta}\mathbf{x}_i - \log \sum_{j=1}^{n} Y_{ij} e^{\boldsymbol{\beta}\mathbf{x}_j} \right]. \tag{5.11}$$

Most partial likelihood programs use some version of the Newton-Raphson algorithm to maximize this function with respect to $\boldsymbol{\beta}$. For details, see Chapter 4 (in the **Maximum Likelihood Estimation: Mathematics** section).

As with PROC LIFEREG, there will occasionally be times when the Newton-Raphson algorithm does not converge. A message in the OUTPUT window will say `WARNING: The information matrix is not positive definite and thus the convergence is questionable.` Unfortunately, PROC PHREG's convergence criterion sometimes makes it look as though the algorithm has converged when, in fact, true convergence is not possible. This problem typically arises when one of the explanatory variables is an indicator variable (1 or 0) and all the observations are censored for one of the levels of that variable. In such cases, the log-likelihood reaches a stable value, but the coefficient of the offending variable keeps heading off toward plus or minus infinity. The only indication of a problem is that the variable in question will have a large coefficient with a much larger standard error. Beginning with SAS 9.2, there is now a simple solution to this problem. Using the FIRTH option in the MODEL statement requests that the partial likelihood function be modified to produce *penalized* partial likelihood estimates. This method reduces the small-sample bias that often occurs with conventional likelihood methods and usually clears up convergence problems (Firth, 1993; Heinze and Schemper, 2001).

To complete the breast cancer example, let's take a look at the partial likelihood results in Output 5.5. With only one covariate, the Global Null Hypothesis statistics provide us with three alternative tests for the effect of that variable. The Wald and score tests have *p*-values that exceed the conventional .05 level, while the likelihood-ratio test has a *p*-value that is slightly below .05. This degree of discrepancy is not at all surprising with a small sample. The estimated hazard ratio of 2.483 tells us that the hazard of death for those whose tumor had the positive marker was nearly 2.5 times the hazard for those without the positive marker.

Because the covariate is dichotomous, an alternative approach is to use PROC LIFETEST to test for differences in survival curves. When I did this, the *p*-value for the log-rank test (.0607) was identical to the *p*-value for the score test in Output 5.5. This is no accident. The log-rank test is the exact equivalent of the partial likelihood score test for a single, dichotomous covariate (when there are no ties).

Output 5.5 *PROC PHREG Results for Breast Cancer Data*

```
              Testing Global Null Hypothesis: BETA=0

         Test                 Chi-Square      DF      Pr > ChiSq

         Likelihood Ratio       3.8843         1        0.0487
         Score                  3.5194         1        0.0607
         Wald                   3.2957         1        0.0695


              Analysis of Maximum Likelihood Estimates

                    Parameter    Standard                               Hazard
Parameter    DF     Estimate      Error    Chi-Square   Pr > ChiSq      Ratio

x            1      0.90933      0.50089     3.2957       0.0695        2.483
```

TIED DATA

The formula for the partial likelihood in equation (5.10) is valid only for data in which no two events occur at the same time. It's quite common for data to contain tied event times, however, so we need an alternative formula to handle those situations. Most partial likelihood programs use a technique called *Breslow's approximation*, which works well when ties are relatively few. But when data are heavily tied, the approximation can be quite poor (Farewell and Prentice, 1980; Hsieh, 1995). Although PROC PHREG uses Breslow's approximation as the default, it also provides a better approximation proposed by Efron (1977) as well as two *exact* methods.

This section explains the background, rationale, and implementation of these alternative methods for handling ties. Because this issue is somewhat confusing, I'm going to discuss it at some length. Those who just want the bottom line can skip to the end of the section where I summarize the practical implications. Because the formulas get rather complicated, I won't go into all the mathematical details. But I will try to provide some intuitive understanding of why there are different approaches and the basic logic of each one.

To illustrate the problem and the various solutions, let's turn again to the recidivism data. As Output 5.6 shows, these data include a substantial number of tied survival times (weeks to first arrest). For weeks 1 through 7, there is only one arrest in each week. For these seven events, the partial likelihood terms are constructed exactly as described in the **Partial Likelihood: Mathematical and Computational Details** section. Five arrests occurred in week 8, however, so the construction of L_8 requires a different method. Two alternative approaches have been proposed for the construction of the likelihood for tied event times; these are specified in PROC PHREG by TIES=EXACT or TIES=DISCRETE as options in the MODEL statement. This terminology is somewhat misleading because both methods give exact partial likelihoods; the difference is that the EXACT method assumes that there is a true but unknown ordering for the tied event times (that is, time is continuous), while the DISCRETE method assumes that the events really occurred at exactly the same time.

Output 5.6 *Week of First Arrest for Recidivism Data*

WEEK	Frequency	Percent	Cumulative Frequency	Cumulative Percent
1	1	0.2	1	0.2
2	1	0.2	2	0.5
3	1	0.2	3	0.7
4	1	0.2	4	0.9
5	1	0.2	5	1.2
6	1	0.2	6	1.4
7	1	0.2	7	1.6
8	5	1.2	12	2.8
9	2	0.5	14	3.2
10	1	0.2	15	3.5
11	2	0.5	17	3.9
12	2	0.5	19	4.4
13	1	0.2	20	4.6
14	3	0.7	23	5.3
15	2	0.5	25	5.8
16	2	0.5	27	6.2
17	3	0.7	30	6.9
18	3	0.7	33	7.6
19	2	0.5	35	8.1
20	5	1.2	40	9.3
21	2	0.5	42	9.7
22	1	0.2	43	10.0
23	1	0.2	44	10.2
24	4	0.9	48	11.1
25	3	0.7	51	11.8
26	3	0.7	54	12.5
27	2	0.5	56	13.0
28	2	0.5	58	13.4
30	2	0.5	60	13.9
31	1	0.2	61	14.1
32	2	0.5	63	14.6
33	2	0.5	65	15.0
34	2	0.5	67	15.5
35	4	0.9	71	16.4
36	3	0.7	74	17.1
37	4	0.9	78	18.1
38	1	0.2	79	18.3
39	2	0.5	81	18.8
40	4	0.9	85	19.7
42	2	0.5	87	20.1
43	4	0.9	91	21.1
44	2	0.5	93	21.5
45	2	0.5	95	22.0
46	4	0.9	99	22.9
47	1	0.2	100	23.1
48	2	0.5	102	23.6
49	5	1.2	107	24.8
50	3	0.7	110	25.5
52	322	74.5	432	100.0

The EXACT Method

Let's begin with the EXACT method because its underlying model is probably more plausible for most applications. Since arrests can occur at any point in time, it's reasonable to suppose that ties are merely the result of imprecise measurement of time and that there is a true but unknown time ordering for the five arrests that occurred in week 8. If we knew that ordering, we could construct the partial likelihood in the usual way. In the absence of any knowledge of that ordering, however, we have to consider all the possibilities. With five events, there are 5! = 120 different possible orderings. Let's denote each of those possibilities by A_i, where $i = 1, \ldots ,$ 120. What we want is the probability of the *union* of those possibilities, that is, $\Pr(A_1$ or A_2 or ... or $A_{120})$. Now, a fundamental law of probability theory is that the probability of the union of a set of mutually exclusive events is just the sum of the probabilities for each of the events. Therefore, we can write

$$L_8 = \sum_{i=1}^{120} \Pr(A_i).$$

(5.12)

Each of these 120 probabilities is just a standard partial likelihood. Suppose, for example, that we arbitrarily label the five arrests at time 8 with the numbers 8, 9, 10, 11, and 12, and suppose further that A_1 denotes the ordering {8, 9, 10, 11, 12}. Then

$$\Pr(A_1) = \left(\frac{e^{\beta x_8}}{e^{\beta x_8} + e^{\beta x_9} + \ldots + e^{\beta x_{432}}} \right) \left(\frac{e^{\beta x_9}}{e^{\beta x_9} + e^{\beta x_{10}} + \ldots + e^{\beta x_{432}}} \right) \cdots \left(\frac{e^{\beta x_{12}}}{e^{\beta x_{12}} + e^{\beta x_{13}} + \ldots + e^{\beta x_{432}}} \right).$$

On the other hand, if A_2 denotes the ordering {9, 8, 10, 11, 12}, we have

$$\Pr(A_2) = \left(\frac{e^{\beta x_9}}{e^{\beta x_8} + e^{\beta x_9} + \ldots + e^{\beta x_{432}}} \right) \left(\frac{e^{\beta x_8}}{e^{\beta x_8} + e^{\beta x_{10}} + \ldots + e^{\beta x_{432}}} \right) \cdots \left(\frac{e^{\beta x_{12}}}{e^{\beta x_{12}} + e^{\beta x_{13}} + \ldots + e^{\beta x_{432}}} \right).$$

We continue in this way for the other 118 possible orderings. Then L_8 is obtained by adding all the probabilities together.

The situation is much simpler for week 9 because only two arrests occurred, giving us two possible orderings. For L_9, then, we have

$$L_9 = \left(\frac{e^{\beta x_{13}}}{e^{\beta x_{13}} + e^{\beta x_{14}} + \ldots + e^{\beta x_{432}}} \right) \left(\frac{e^{\beta x_{14}}}{e^{\beta x_{14}} + e^{\beta x_{15}} + \ldots + e^{\beta x_{432}}} \right) +$$

$$\left(\frac{e^{\beta x_{14}}}{e^{\beta x_{13}} + e^{\beta x_{14}} + \ldots + e^{\beta x_{432}}} \right) \left(\frac{e^{\beta x_{13}}}{e^{\beta x_{13}} + e^{\beta x_{15}} + \ldots + e^{\beta x_{432}}} \right)$$

where the numbers 13 and 14 are arbitrarily assigned to the two events. When we get to week 10, there's only one event, so we're back to the standard partial likelihood formula:

$$L_{10} = \left(\frac{e^{\beta x_{15}}}{e^{\beta x_{15}} + e^{\beta x_{16}} + \ldots + e^{\beta x_{432}}} \right).$$

It's difficult to write a general formula for the exact likelihood with tied data because the notation becomes very cumbersome. For one version of a general formula, see Kalbfleisch and Prentice (2002). Be forewarned that the formula in the official PROC PHREG documentation bears no resemblance to that given by Kalbfleisch and Prentice or to the explanation given here. That's because it's based on a re-expression of the formula in terms of a definite integral, which facilitates computation (DeLong, Guirguis, and So, 1994).

It should be obvious, by this point, that computation of the exact likelihood can be a daunting task. With just five tied survival times, we have seen that one portion of the partial likelihood increased from 1 term to 120 terms. If 10 events occur at the same time, there are over 3 million possible orderings to evaluate. Until recently, statisticians abandoned all hope that such computations might be practical. What makes it possible now is the development of an integral representation of the likelihood, which is much easier to evaluate numerically (DeLong, Guirguis, and So, 1994). Even with this innovation, however, computation of the exact likelihood when large numbers of events occur at the same time can take a lot of computing time.

Early recognition of these computational difficulties led to the development of approximations. The most popular of these is widely attributed to Breslow (1974), but it was first proposed by Peto (1972). This is the default in PROC PHREG, and it is nearly universal in other programs. Efron (1977) proposed an alternative approximation that is also available in PROC PHREG. The results that we saw earlier in Output 5.1 for the recidivism data were obtained with the Breslow approximation.

To use the EXACT method, we specify

```
PROC PHREG DATA=recid;
   MODEL week*arrest(0)=fin age race wexp mar paro prio
         / TIES=EXACT;
RUN;
```

This PROC step produces the results in Output 5.7. Comparing this with Output 5.1, it's apparent that the Breslow approximation works well in this case. The coefficients are generally the same to at least two (and sometimes three) decimal places. The test statistics all yield the same conclusions.

Output 5.7 *Recidivism Results Using the EXACT Method*

```
                Testing Global Null Hypothesis: BETA=0

        Test                  Chi-Square      DF     Pr > ChiSq

        Likelihood Ratio        33.2663        7      <.0001
        Score                   33.5289        7      <.0001
        Wald                    32.1190        7      <.0001


              Analysis of Maximum Likelihood Estimates

                   Parameter   Standard                          Hazard
 Parameter   DF     Estimate      Error   Chi-Square  Pr > ChiSq   Ratio

 fin          1     -0.37942    0.19138     3.9305      0.0474     0.684
 age          1     -0.05743    0.02200     6.8152      0.0090     0.944
 race         1      0.31393    0.30800     1.0389      0.3081     1.369
 wexp         1     -0.14981    0.21223     0.4983      0.4803     0.861
 mar          1     -0.43372    0.38187     1.2900      0.2560     0.648
 paro         1     -0.08486    0.19576     0.1879      0.6646     0.919
 prio         1      0.09152    0.02865    10.2067      0.0014     1.096
```

Output 5.8 shows the results from using Efron's approximation (invoked by using TIES=EFRON). If Breslow's approximation is good, this one is superb. Nearly all the numbers are the same to four decimal places. In all cases where I've tried the two approximations, Efron's approximation gave results that were much closer to the exact results than Breslow's approximation. This improvement comes with only a trivial increase in computation time.

Output 5.8 *Recidivism Results Using Efron's Approximation*

```
                    Testing Global Null Hypothesis: BETA=0

         Test                   Chi-Square      DF      Pr > ChiSq

         Likelihood Ratio        33.2659         7        <.0001
         Score                   33.5287         7        <.0001
         Wald                    32.1192         7        <.0001


                    Analysis of Maximum Likelihood Estimates

                         Parameter    Standard                              Hazard
         Parameter   DF    Estimate      Error   Chi-Square   Pr > ChiSq     Ratio

         fin          1    -0.37942    0.19138     3.9304       0.0474       0.684
         age          1    -0.05743    0.02200     6.8152       0.0090       0.944
         race         1     0.31392    0.30799     1.0389       0.3081       1.369
         wexp         1    -0.14981    0.21223     0.4983       0.4803       0.861
         mar          1    -0.43372    0.38187     1.2900       0.2560       0.648
         paro         1    -0.08486    0.19576     0.1879       0.6646       0.919
         prio         1     0.09152    0.02865    10.2067       0.0014       1.096
```

If the approximations are so good, why do we need the computationally intensive EXACT method? Farewell and Prentice (1980) showed that the Breslow approximation deteriorates as the number of ties at a particular point in time becomes a large proportion of the number of cases at risk. For the recidivism data in Output 5.6, the number of tied survival times at any given time point is never larger than 2 percent of the number at risk, so it's not surprising that the approximations work well.

Now let's look at an example where the conditions are less favorable. The data consist of 100 simulated job durations, measured from the year of entry into the job until the year that the employee quit. Durations after the fifth year are censored. If the employee was fired before the fifth year, the duration is censored at the end of the last full year in which the employee was working. We know only the year in which the employee quit, so the survival times have values of 1, 2, 3, 4, or 5.

Here's a simple life table for these data:

Duration	Number Quit	Number Censored	Number At Risk	Number Quit/ At Risk
1	22	7	100	.22
2	18	3	71	.25
3	16	4	50	.32
4	8	1	30	.27
5	4	17	21	.19

The number at risk at each duration is equal to the total number of cases (100) minus the number who quit or were censored at previous durations. Looking at the last column, we see that the ratio of the number quitting to the number at risk is substantial at each of the five points in time. Three covariates were measured at the beginning of the job: years of schooling (ED), salary in thousands of dollars (SALARY), and prestige of the occupation (PRESTIGE) measured on a scale from 1 to 100.

Output 5.9 displays selected results from using PROC PHREG with the three different methods for handling ties. Breslow's method yields coefficient estimates that are about one-third smaller in magnitude than those using the EXACT method, while the *p*-values (for testing the hypothesis that each coefficient is 0) are substantially higher. In fact, the *p*-value for the SALARY variable is above the .05 level for Breslow's method, but it is only .01 for the EXACT method. Efron's method produces coefficients that are about midway between the other two methods, but the *p*-values are much closer to those of the EXACT method. Clearly, the Breslow approximation is unacceptable for this application. Efron's approximation is not bad for drawing qualitative conclusions, but there is an appreciable loss of accuracy in estimating the magnitudes of the coefficients.

Output 5.9 *Results for Job Duration Data: Three Methods for Handling Ties*

```
Ties Handling: BRESLOW

              Parameter   Standard    Wald      Pr >     Hazard
  Variable DF  Estimate     Error  Chi-Square Chi-Square  Ratio

  ED       1    0.116453   0.05918   3.87257    0.0491    1.124
  PRESTIGE 1   -0.064278   0.00959  44.93725    0.0001    0.938
  SALARY   1   -0.014957   0.00792   3.56573    0.0590    0.985


Ties Handling: EFRON

              Parameter   Standard    Wald      Pr >     Hazard
  Variable DF  Estimate     Error  Chi-Square Chi-Square  Ratio

  ED       1    0.144044   0.05954   5.85271    0.0156    1.155
  PRESTIGE 1   -0.079807   0.00996  64.20009    0.0001    0.923
  SALARY   1   -0.020159   0.00830   5.90363    0.0151    0.980


Ties Handling: EXACT

              Parameter   Standard    Wald      Pr >     Hazard
  Variable DF  Estimate     Error  Chi-Square Chi-Square  Ratio

  ED       1    0.164332   0.06380   6.63419    0.0100    1.179
  PRESTIGE 1   -0.092019   0.01240  55.10969    0.0001    0.912
  SALARY   1   -0.022545   0.00884   6.50490    0.0108    0.978
```

The DISCRETE Method

The DISCRETE option in PROC PHREG is also an exact method but one based on a fundamentally different model. In fact, it is not a proportional hazards model at all. The model does fall within the framework of Cox regression, however, because it was proposed by Cox in his original 1972 paper and because the estimation method is a form of partial likelihood. Unlike the EXACT model, which assumes that ties are merely the result of imprecise measurement of time, the DISCRETE model assumes that time is really discrete. When two or more events appear to happen at the same time, there is no underlying ordering—they really happen at the same time.

While most applications of survival analysis involve events that can occur at any moment on the time continuum, there are definitely some events that are best treated as if time were discrete. If the event of interest is a change in the political party occupying the U.S. presidency, that can only occur once every four years. Or suppose the aim is to predict how many months it takes before a new homeowner misses a mortgage payment. Because payments are only due at monthly intervals, a discrete-time model is the natural way to go.

Cox's model for discrete-time data can be described as follows. The time variable t can only take on integer values. Let P_{it} be the conditional probability that individual i has an event at time t, given that an event has not already occurred to that individual. This probability is sometimes called the *discrete-time hazard*. The model says that P_{it} is related to the covariates by a logistic regression equation:

$$\log\left(\frac{P_{it}}{1 - P_{it}}\right) = \alpha_t + \beta_1 x_{i1} + \ldots + \beta_k x_{ik}.$$

The expression on the left side of the equation is the logit or log-odds of P_{it}. On the right side, we have a linear function of the covariates, plus a term α_t that plays the same role as $\alpha(t)$ in expression (5.2) for the proportional hazards model. α_t is just a set of constants—one for each time point—that can vary arbitrarily from one time point to another.

This model can be described as a proportional *odds* model. The odds that individual i has an event at time t (given that i did not already have an event) is $O_{it} = P_{it} / (1 - P_{it})$. The model implies that the ratio of the odds for any two individuals O_{it} / O_{jt} does *not* depend on time (although it may vary with the covariates).

How can we estimate this model? In Chapter 7, "Analysis of Tied or Discrete Data with PROC LOGISTIC," we will see how to estimate it using standard maximum likelihood methods that yield estimates of both the β coefficients and the α_ts. Using partial likelihood, however, we can treat the α_ts as nuisance parameters and estimate only the βs. If there are J unique times at which events occur, there will be J terms in the partial likelihood function:

$$PL = \prod_{j=1}^{J} L_j$$

where L_j is the partial likelihood of the jth event. Thus, for the job duration data, there are only five terms in the partial likelihood function. But each of those five terms is colossal. Here's why.

At time 1, 22 people had events out of 100 people who were at risk. To get L_1, we ask the question: given that 22 events occurred, what is the probability that they occurred to these particular 22 people rather than to some different set of 22 people from among the 100 at risk? How many different ways are there of selecting 22 people from among a set of 100? A lot! Specifically, 7.3321×10^{21}. Let's call that number Q, and let q be a running index from 1 to Q, with $q = 1$ denoting the set that actually experienced the events. For a given set q, let ψ_q be the product of the odds for all the individuals in that set. Thus, if the individuals who actually experienced events are labeled $i = 1$ to 22, we have

$$\psi_1 = \prod_{i=1}^{22} O_{i1} \, .$$

We can then write

$$L_1 = \frac{\psi_1}{\psi_1 + \psi_2 \ldots + \psi_Q} \, .$$

This looks like a simple expression, but there are *trillions* of terms being summed in the denominator. Fortunately, there is a recursive algorithm that makes it practical, even with substantial numbers of ties (Gail et al., 1981). Still, doing this with a very large data set with many ties can take a lot of computer time.

Does the discrete-time model make sense for these data? For most jobs it's possible to quit at any point in time, suggesting that the model might not be appropriate. Remember, however, that these are simulated data. Because the simulation is actually based on a discrete-time model, it makes perfectly good sense in this case. Output 5.10 displays the results. Comparing these with the results for the EXACT method in Output 5.9, we see that the chi-square statistics and the p-values are similar. However, the coefficients for the DISCRETE method are about one-third larger for ED and PRESTIGE and about 15 percent larger for SALARY. This discrepancy is due largely to the fact that completely different models are being estimated, a hazard model and a logit model. The logit coefficients will usually be larger. For the logit model, $100(e^\beta - 1)$ gives the percent change in the *odds* that an event will occur for a one-unit increase in the covariate. Thus, each additional year of schooling increases the odds of quitting a job by $100(e^{.219}-1) = 24$ percent.

Output 5.10   *Job Duration Results Using the DISCRETE Method*

```
Ties Handling: DISCRETE

              Analysis of Maximum Likelihood Estimates

                  Parameter   Standard    Wald       Pr >      Hazard
    Variable DF    Estimate     Error   Chi-Square Chi-Square   Ratio

    ED        1    0.219378    0.08480    6.69295    0.0097     1.245
    PRESTIGE  1   -0.120474    0.01776   46.02220    0.0001     0.886
    SALARY    1   -0.026108    0.01020    6.55603    0.0105     0.974
```

Comparison of Methods

Though the job duration coefficients differ for the two exact methods, they are at least in the same ballpark. More generally, it has been shown that if ties result from grouping continuous time data into intervals, the logit model converges to the proportional hazards model as the interval length gets smaller (Thompson, 1977). When there are no ties, the partial likelihoods for all four methods (the two exact methods and the two approximations) reduce to the same formula.

The examples that we've seen so far have been small enough, both in number of observations and numbers of ties, that the computing times for the two exact methods were quite small (much less than a second). Before concluding, let's see what happens with larger data sets. I took the 100 observations in the job duration data set and duplicated them to produce data sets of size 1000 to 6000 in increments of 1000. I then estimated the models on a Dell laptop running at 1.8GHz with 1 Gb of memory. Here are the CPU times, in seconds, for the two exact methods and six sample sizes:

	1000	2000	3000	4000	5000	6000
EXACT	3.6	16	43	93	171	316
DISCRETE	1.6	6	13	23	33	51

The good news is that these computing times are more than an order of magnitude smaller than the times that I reported in the 1995 edition of this book. For 1000 observations, what previously took the EXACT method more than 2 minutes now takes less than 4 seconds. The upshot is that these methods are now practical for fairly large data sets, at least if you are willing to be patient. For the DISCRETE method, computing time goes up approximately with the square of the sample size. Computing time for the EXACT method goes up even faster. Incidentally, both approximate methods (BRESLOW and EFRON) took less than a quarter of a second for 6000 observations.

What we've learned about the handling of ties can be summarized in six points:

- When there are no ties, all four options in PROC PHREG give identical results.
- When there are few ties, it makes little difference which method is used. But because computing times will also be comparable, you might as well use one of the exact methods.
- When the number of ties is large, relative to the number at risk, the approximate methods tend to yield coefficients that are biased toward 0.

- Both the EXACT and DISCRETE methods produce exact results (that is, true partial likelihood estimates), but the EXACT method assumes that ties arise from grouping continuous, untied data, while the DISCRETE method assumes that events really occur at the same, discrete times. The choice should be based on substantive grounds, although qualitative results will usually be similar.
- Both of the exact methods need a substantial amount of computer time for large data sets containing many ties, especially the EXACT method.
- If the exact methods are too time-consuming, use the Efron approximation, at least for model exploration. It's nearly always better than the Breslow method, with virtually no increase in computer time.

TIME-DEPENDENT COVARIATES

Time-dependent covariates are those that may change in value over the course of observation. While it's simple to modify Cox's model to allow for time-dependent covariates, the computation of the resulting partial likelihood is more time-consuming, and the practical issues surrounding the implementation of the procedure can be quite complex. It's easy to make mistakes without realizing it, so be sure you know what you're doing.

To modify the model in equation (5.2) to include time-dependent covariates, all we need to do is write (t) after the xs that are time-dependent. For a model with one time-invariant covariate and one time-dependent covariate, we have

$$\log h_i(t) = \alpha(t) + \beta_1 x_{i1} + \beta_2 x_{i2}(t).$$

This says that the hazard at time t depends on the value of x_1 and on the value of x_2 at time t. What may not be clear is that $x_2(t)$ can be defined using any information about the individual before time t, thereby allowing for lagged or cumulative values of some variables. For example, if we want a model in which the hazard of arrest depends on employment status, we could specify employment as

- whether the person is currently employed in week t
- whether the person was employed in the previous month (t-1)
- the number of weeks of employment in the preceding 3 months
- the number of bouts of unemployment in the preceding 12 months.

The use of lagged covariates is often essential for resolving issues of causal ordering (more on that later).

For handling time-dependent covariates, PROC PHREG has two very different ways of setting up the data and specifying the model. In the *programming statements* method, there is one record per individual, and the information about time-dependent covariates is contained in multiple variables on that record. The time-dependent covariates are then defined in programming statements that are part of the PROC PHREG step. In the *counting process* method, on the other hand, there may be multiple records for each individual, with each record corresponding to an interval of time during which all the covariates remain constant. Once this special data set has been constructed, time-dependent covariates are treated just like time-invariant covariates. However, a different syntax (the counting process syntax) is needed to specify the dependent variable.

If done properly, these two methods produce exactly the same results, so the choice between them is purely a matter of computational convenience. Both methods may require a substantial amount of careful programming. However, with the counting process syntax, all the programming is up front in the construction of the data set. That makes it straightforward to check the data set for any programming errors. The method also lends itself to a division of labor in which a skilled programmer produces the required data set. The data analyst can then easily specify whatever models are desired in PROC PHREG.

With the programming statements method, on the other hand, the program to generate the time-dependent covariates must be a part of every PROC PHREG run. Furthermore, because the temporary data set produced by the program is not accessible, there is no easy way to check for programming mistakes. All this would seem to favor the counting process method. Nonetheless, as we shall see, the counting process method often requires considerably more program code. We will illustrate these two methods with several examples, first applying the programming statements method and then using the counting process method.

Heart Transplant Example

We begin with another look at the Stanford Heart Transplant Data. In the **Partial Likelihood: Examples** section, we attempted to determine whether a transplant raised or lowered the risk of death by examining the effect of a time-invariant covariate TRANS that was equal to 1 if the patient ever had a transplant and 0 otherwise. I claimed that those results were completely misleading because patients who died quickly had less time available to get transplants. Now we'll do it right by defining a time-dependent covariate PLANT equal to 1 if the patient has already had a transplant at day *t*; otherwise, PLANT is equal to 0.

Here's how it's done with the programming statements method:

```
PROC PHREG DATA=stan;
   MODEL surv1*dead(0)=plant surg ageaccpt / TIES=EFRON;
   IF wait>=surv1 OR wait=. THEN plant=0; ELSE plant=1;
RUN;
```

Recall that SURV1 is the time in days from acceptance into the program until death or termination of observation (censoring). Recall also that SURG =1 if the patient had previous heart surgery; otherwise, SURG=0. AGEACCPT is the patient's age in years at the time of acceptance, and WAIT is the time in days from acceptance until transplant surgery, coded as missing for those who did not receive transplants. Note that I have used the EFRON method for handling ties. Although there are not many ties in these data, the discussion in the last section concluded that the EFRON method is a better default than the usual BRESLOW method.

The IF statement defines the new time-varying covariate PLANT. Note that programming statements must *follow* the MODEL statement. At first glance, this IF statement may be puzzling. For patients who were not transplanted, the IF condition will be true (and PLANT will be assigned a value of 0) because their WAIT value will be missing. On the other hand, for those who did receive transplants, it appears that the IF condition will always be false because their waiting time to transplant cannot be greater than their survival time, giving us a fixed covariate rather than a time-dependent covariate. Now it's true that waiting time is never greater than survival time for transplanted patients (with one patient dying on the operating table). But, unlike an IF statement in the DATA step, which only operates on a single case at a time, this IF statement compares waiting times for patients who were at risk of a death with survival times for patients who experienced events. Thus, SURV1 in this statement is typically not the patient's own survival time, but the survival time of some other patient who died. This fact will become clearer in the next subsection when we examine the construction of the partial likelihood.

Results in Output 5.11 indicate that transplantation has no effect on the hazard of death. The effect of age at acceptance is somewhat smaller than it was in Output 5.2, although still statistically significant. However, the effect of prior heart surgery is larger and now significant at the .05 level.

Output 5.11   *Results for Transplant Data with a Time-Dependent Covariate*

Parameter	DF	Parameter Estimate	Standard Error	Chi-Square	Pr > ChiSq	Hazard Ratio
plant	1	0.01610	0.30859	0.0027	0.9584	1.016
surg	1	-0.77332	0.35967	4.6229	0.0315	0.461
ageaccpt	1	0.03054	0.01389	4.8312	0.0279	1.031

Now let's redo the analysis using the counting process method. To do this, we must construct a data set with multiple records per person, one record for each period of time during which all the covariates remain constant. For persons who did not receive transplants, only one record is needed. Persons who received transplants need two records: one for the interval of time between acceptance and transplantation (with PLANT=0) and a second record for the interval between transplantation and either death or censoring (PLANT=1). The first interval is coded as censored. The second interval may be either censored or uncensored, depending on whether the person died during the observation period. Time-invariant variables are replicated for the two records. For each record, we also need two time variables, one for the start of the interval and one for the end of the interval. Both time variables are measured in days since acceptance into the study.

Here is the DATA step needed to produce the required data set:

```
DATA stanlong;
SET stan;
plant=0;
start=0;
IF trans=0 THEN DO;
  dead2=dead;
  stop=surv1;
  IF stop=0 THEN stop=.1;
  OUTPUT;
END;
ELSE DO;
  stop=wait;
  IF stop=0 THEN stop=.1;
  dead2=0;
  OUTPUT;
  plant=1;
  start=wait;
  IF stop=.1 THEN start=.1;
  stop=surv1;
  dead2=dead;
  OUTPUT;
END;
RUN;
```

Clearly this is a lot more programming than was required for the programming statements method. Here is a brief explanation. The first two assignment statements initialize the time-dependent covariate (PLANT) and the starting time at 0. The IF-DO statement begins a block of definitions for those who did not receive transplants. A new censoring indicator, DEAD2, is specified to have the same value as the old indicator DEAD, and the stop time is set to be the same as the original time variable SURV1. The OUTPUT statement writes the record to the new data set. By default, all other variables in the original data set are included in this record.

A peculiarity of this data set is that there were two individuals who died on their day of acceptance into the study, so both their start time and their stop time would be 0. Unfortunately, the counting process syntax deletes records that have the same start and stop time. That's why there is an IF statement that changes a stop time of 0 to .1.

The ELSE DO statement begins a block of variable definitions for those who received transplants. First, a record is created for the interval from acceptance to transplant. The stop time is set to be the waiting time to transplant. Again, because there was one person who was transplanted on the day of acceptance, it is necessary to change the stop time from 0 to .1. The censoring indicator DEAD2 is set to 0 since this interval did not end in a death. And the OUTPUT statement writes the record to the new data set.

Next, a record is created for the interval from transplant to death or censoring. PLANT is now set to 1. The START time is set to the time of transplant (WAIT). The STOP time is reset to the original death or censoring time, and the censoring indicator is reset to its original values. For the one person transplanted on the day of acceptance, the IF statement aligns the start time of the second interval with the stop time of the first interval.

The new data set has 172 records, one each for the 34 people who did not receive transplants and two records for each of the 69 people who did receive transplants. Output 5.12 displays the last 15 records in the new data set, representing eight people. All but one (record 164) had a transplant and, therefore, the rest have two records. Only one person died (record 163).

Output 5.12 *Last 15 Records for Counting Process Data Set*

Obs	start	stop	dead2	plant	surg	ageaccpt
158	0.0	35.0	0	0	1	36.6543
159	35.0	1141.0	0	1	1	36.6543
160	0.0	57.0	0	0	1	45.3032
161	57.0	1321.0	0	1	1	45.3032
162	0.0	36.0	0	0	0	54.0123
163	36.0	1386.0	1	1	0	54.0123
164	0.0	1400.0	0	0	0	30.5352
165	0.0	40.0	0	0	1	48.4819
166	40.0	1407.0	0	1	1	48.4819
167	0.0	22.0	0	0	0	40.5530
168	22.0	1571.0	0	1	0	40.5530
169	0.0	50.0	0	0	0	48.9035
170	50.0	1586.0	0	1	0	48.9035
171	0.0	24.0	0	0	0	33.2238
172	24.0	1799.0	0	1	0	33.2238

To estimate the model, we use the counting process syntax, specifying both the starting time and the stopping time for each record:

```
PROC PHREG DATA=stanlong;
   MODEL (start,stop)*dead2(0)=plant surg ageaccpt /TIES=EFRON;
RUN;
```

Results are identical to those in Output 5.11.

Although this syntax is similar to the PROC LIFEREG syntax for interval-censored data, the intent is completely different. In the PROC LIFEREG syntax, the event is known to occur sometime within the specified interval. In the counting process syntax, on the other hand, (START, STOP) represents an interval of time during which the individual was continuously at risk of the event, and events can only occur at the end of the interval.

Construction of the Partial Likelihood with Time-Dependent Covariates

With time-dependent covariates, the partial likelihood function has the same form that we saw previously in equation (5.4) and equation (5.10). The only thing that changes is that the covariates are now indexed by time. Consider, for example, the 12 selected cases from the Stanford Heart Transplant Data that are shown in Output 5.13. These are all the cases that had death or censoring times between 16 and 38 days, inclusive.

Output 5.13 *Selected Cases from the Stanford Heart Transplant Data*

OBS	SURV1	DEAD	WAIT
19	16	1	4
20	17	1	.
21	20	1	.
22	20	1	.
23	27	1	17
24	29	1	4
25	30	0	.
26	31	1	.
27	34	1	.
28	35	1	.
29	36	1	.
30	38	1	35

On day 16, one death occurred (to case 19). Because 18 people had already died or been censored by day 16, there were 103–18 = 85 people left in the risk set on that day. Let's suppose that we have a single covariate, the time-dependent version of transplant status. The partial likelihood for day 16 is therefore

$$\frac{e^{\beta x_{19}(16)}}{e^{\beta x_{19}(16)} + e^{\beta x_{20}(16)} + e^{\beta x_{21}(16)} + \ldots + e^{\beta x_{103}(16)}}.$$

To calculate this quantity, PROC PHREG must compute the value of x on day 16 for each of the 85 people at risk. For cases 19 and 24, a transplant occurred on day 4. Since this was before day 16, we have $x_{19}(16) = 1$ and $x_{24}(16) = 1$. Waiting time is missing for cases 20–22 and cases 25–29, indicating that they never received a transplant. Therefore, $x_{20}(16) = x_{21}(16) = x_{22}(16) = x_{25}(16) = x_{26}(16) = x_{27}(16) = x_{28}(16) = x_{29}(16) = 0$. Case 23 had a transplant on day 17, so on day 16, the patient was still without a transplant and $x_{23}(16) = 0$. Similarly, we have $x_{30}(16) = 0$ because case 30 didn't get a transplant until day 35.

For the programming statements method, the calculation of the appropriate x values is accomplished by the IF statement shown earlier in the heart transplant example. At each unique event time, PROC PHREG calculates a term in the partial likelihood function (like the one above) by applying the IF statement to all the cases in the risk set at that time. Again, the value of SURV1 in the IF statement is the event time that PROC PHREG is currently operating on, not the survival time for each individual at risk.

For the counting process method, to compute the partial likelihood for a death occurring on day 16, PHREG simply searches through all the records in the data set to find those with START and STOP times that bracket day 16. Note that the intervals defined by those times are open on the left and closed on the right. That means that, if the START time is 16 and the STOP time is 24, day 24 would be included in the interval but day 16 would not be included.

This property of the counting process method has important implications for the next term in the partial likelihood function, which corresponds to the death that occurred to person 20 on day 17:

$$\frac{e^{\beta x_{20}(17)}}{e^{\beta x_{20}(17)} + e^{\beta x_{21}(17)} + \ldots + e^{\beta x_{103}(17)}}.$$

Of course person 19 no longer shows up in this formula because that patient left the risk set at death. The values of x are all the same as they were for day 16, even for patient 23, who had a transplant on day 17. This patient has two records, one running from 0 to 17 and the other from 17 to 27. The first record includes day 17, but the second record does not. The first record has PLANT=0, so $x_{23}(17) = 0$.

For the data in Output 5.13, there are eight additional terms in the partial likelihood function. For the first five of these, the values of x for cases remaining in the risk set are the same as they were on day 17. On day 36, however, the value of x for case 30 switches from 0 to 1.

Covariates Representing Alternative Time Origins

When I discussed the choice of time origin in Chapter 2, I mentioned that you can include alternative origins as covariates, sometimes as time-dependent covariates. Let's see how this might work with the Stanford Heart Transplant Data. In the analysis just completed, the origin was the date of acceptance into the program, with the time of death or censoring computed from that point. It is certainly plausible, however, that the hazard of death also depends on age or calendar time. We have already included age at acceptance into the program as a time-constant covariate and found that patients who were older at acceptance had a higher hazard of death. But if that's the case, we might also expect that the hazard will continue to increase with age *after* acceptance into the program. A natural way to allow for this possibility is to specify a model with *current* age as a time-dependent covariate. Here's how to do that with PROC PHREG:

```
PROC PHREG DATA=stan;
   MODEL surv1*dead(0)=plant surg age / TIES=EFRON;
   IF wait>surv1 OR wait=. THEN plant=0; ELSE plant=1;
   age=ageaccpt+surv1;
RUN;
```

In this program, current age is defined as age at acceptance plus the time to the current event. While this is certainly correct, a surprising thing happens: the results are *exactly* the same as in Output 5.11. Here's why. We can write the time-dependent version of the model as

$$\log h(t) = \alpha(t) + \beta_1 x_1 + \beta_2 x_2(t) + \beta_3 x_3(t) \tag{5.13}$$

where x_1 is the surgery indicator, x_2 is the transplant status, and x_3 is the current age. We also know that $x_3(t) = x_3(0) + t$, where $x_3(0)$ is age at the time of acceptance. Substituting into equation (5.13), we have

$$\log h(t) = \alpha^*(t) + \beta_1 x_1 + \beta_2 x_2(t) + \beta_3 x_3(0)$$

where $\alpha^*(t) = \alpha(t) + \beta_3 t$. Thus, we have converted a model with a time-dependent version of age to one with a time-invariant version of age. In the process, the arbitrary function of time changes, but that's of no consequence because it drops out of the estimating equations anyway. The same trick works with calendar time: instead of specifying a model in which the hazard depends on current calendar time, we can estimate a model with calendar time at the point of acceptance and get exactly the same results.

The trick does not work, however, if the model says that the log of the hazard is a *non*linear function of the alternative time origin. For example, suppose we want to estimate the model

$$\log h(t) = \alpha(t) + \beta_1 x_1 + \beta_2 x_2(t) + \beta_3 \log x_3(t)$$

where $x_3(t)$ is again age at time t. Substitution with $x_3(t) = x_3(0) + t$ gets us nowhere in this case because the β_3 coefficient does not distribute across the two components of $x_3(t)$. You must estimate this model with $\log x_3(t)$ as a time-dependent covariate:

```
PROC PHREG DATA=stan;
   MODEL surv1*dead(0)=plant surg logage / TIES=EFRON;
   IF wait>surv1 OR wait=. THEN plant=0; ELSE plant=1;
   logage=LOG(ageaccpt+surv1);
RUN;
```

In sum, if you are willing to forego nonlinear functions of time, you can include any alternative time origin as a time-invariant covariate, measured at the origin that is actually used in calculating event times.

Time-Dependent Covariates Measured at Regular Intervals

As we saw earlier, calculation of the partial likelihood requires that the values of the covariates be known for every individual who was at risk at each event time. In practice, because we never know in advance when events will occur, we need to know the values of all the time-dependent covariates at every point in time. This requirement was met for the heart transplant data: death times were measured in days, and for each day, we could construct a variable indicating whether a given patient had already had a heart transplant.

Often, however, the information about the covariates is only collected at regular intervals of time that may be longer (or shorter) than the time units used to measure event times. For example, in a study of time to death among AIDS patients, there may be monthly follow-ups in which vital signs and blood measurements are taken. If deaths are reported in days, there is only about a 1-in-30 chance that the time-dependent measurements will be available for a given patient on a particular death day. In such cases, it is necessary to use some ad-hoc method for assigning covariate values to death days. In a moment, I will discuss several issues related to such ad-hoc approaches. First, let's look at an example in which the time intervals for covariate measurement correspond exactly to the intervals in which event times are measured.

For the recidivism example that we studied earlier, additional information was available on the employment status of the released convicts over the 1-year follow-up period. Specifically, for each of the 52 weeks of follow-up, there was a dummy variable coded 1 if the person was employed full-time during that week; otherwise, the variable was coded 0. The data are read as follows:

```
DATA RECID;
   INFILE 'c:\recid.dat';
   INPUT week arrest fin age race wexp mar paro prio educ emp1-emp52;
RUN;
```

The employment status information is contained in the variables EMP1 to EMP52. For the programming statements method, we can work directly with the RECID data set. The PROC PHREG statements are

```
PROC PHREG DATA=recid;
   MODEL week*arrest(0)=fin age race wexp mar paro prio employed
       / TIES=EFRON;
   ARRAY emp(*) emp1-emp52;
   DO i=1 TO 52;
      IF week=i THEN employed=emp[i];
   END;
RUN;
```

The aim here is to pick out the employment indicator that corresponds to the week in which an event occurred and assign that value to the variable EMPLOYED for everyone who is in the risk set for that week. The ARRAY statement makes it possible to treat the 52 distinct dummy variables as a single subscripted array, thereby greatly facilitating the subsequent operations. Note that in the IF statement, the "subscript" to EMP must be contained in brackets rather than parentheses.

The only problem with this code is that the program has to cycle through 52 IF statements to pick out the right value of the employment variable. A more efficient program that directly retrieves the correct value is as follows:

```
PROC PHREG DATA=recid;
    MODEL week*arrest(0)=fin age race wexp mar paro prio employed
        / TIES=EFRON;
    ARRAY emp(*) emp1-emp52;
    employed=emp[week];
RUN;
```

Output 5.14 shows the results (for either version). For the time-invariant variables, the coefficients and test statistics are pretty much the same as in Output 5.8. Judging by the chi-square test, however, the new variable EMPLOYED has by far the strongest effect of any variable in the model. The hazard ratio of .265 tells us that the hazard of arrest for those who were employed full time is a little more than one-fourth the hazard for those who were not employed full time.

Output 5.14 *Recidivism Results with a Time-Dependent Covariate*

Parameter	DF	Parameter Estimate	Standard Error	Chi-Square	Pr > ChiSq	Hazard Ratio
fin	1	-0.35672	0.19113	3.4835	0.0620	0.700
age	1	-0.04633	0.02174	4.5442	0.0330	0.955
race	1	0.33867	0.30960	1.1966	0.2740	1.403
wexp	1	-0.02557	0.21142	0.0146	0.9037	0.975
mar	1	-0.29374	0.38303	0.5881	0.4432	0.745
paro	1	-0.06420	0.19468	0.1088	0.7416	0.938
prio	1	0.08515	0.02896	8.6455	0.0033	1.089
employed	1	-1.32823	0.25071	28.0679	<.0001	0.265

Unfortunately, these results are undermined by the possibility that arrests affect employment status rather than vice versa. If someone is arrested and incarcerated near the beginning of a particular week, the probability of working full time during the remainder of that week is likely to drop precipitously. This potential reverse causation is a problem that is quite common with time-dependent covariates, especially when event times or covariate times are not measured precisely.

One way to reduce ambiguity in the causal ordering is to *lag* the covariate values. Instead of predicting arrests in a given week based on employment status in the same week, we can use employment status in the prior week. This requires only minor modifications in the SAS code:

```
PROC PHREG;
   WHERE week>1;
   MODEL week*arrest(0)=fin age race wexp mar paro prio employed
         / TIES=EFRON;
   ARRAY emp(*) emp1-emp52;
   employed=emp[week-1];
RUN;
```

One change is to add a WHERE statement to eliminate cases (only one case, in fact) with an arrest in the first week after release. This change is necessary because there were no values of employment status before the first week. The other change is to use WEEK–1 as the subscript for EMP rather than WEEK. With these changes, the coefficient for EMPLOYED drops substantially, from −1.33 to −.79, which implies that the hazard of arrest for those who were employed is about 45 percent of the risk of those who were not employed. While this is a weaker effect than we found using unlagged values of employment status, it is still highly significant with a chi-square value of 13.1. The effects of the other variables remain virtually unchanged.

As this example points out, there are often many different ways of specifying the effect of a time-dependent covariate. Let's consider a couple of the alternatives. Instead of a single lagged version of the employment status indicator, we can have both a 1-week and a 2-week lag, as shown in the following program:

```
PROC PHREG;
   WHERE week>2;
   MODEL week*arrest(0)=fin age race wexp mar paro prio employ1
         employ2 / TIES=EFRON;
   ARRAY emp(*) emp1-emp52;
   employ1=emp[week-1];
   employ2=emp[week-2];
RUN;
```

Note that because of the 2-week lag, it is necessary to eliminate cases with events in either week 1 or week 2. When I tried this variation, I found that *neither* EMPLOY1 nor EMPLOY2 was significant (probably because they are highly correlated—a joint test that both coefficients are 0 yielded a *p*-value of .0018). However, the 1-week lag was much stronger than the 2-week lag. So it looks as though we're better off sticking with the single 1-week lag.

Another possibility is that the hazard of arrest may depend on the *cumulative* employment experience after release rather than the employment status in the preceding week. Consider the following SAS code:

```
DATA RECIDCUM;
   SET recid;
   ARRAY emp(*) emp1-emp52;
   ARRAY cum(*) cum1-cum52;
   cum1=emp1;
   DO i=2 TO 52;
      cum(i)=(cum(i-1)*(i-1) + emp(i))/i;
   END;
PROC PHREG DATA=recidcum;
   WHERE week>1;
   MODEL week*arrest(0)=fin age race wexp mar paro prio employ
         / TIES=EFRON;
   ARRAY cumemp(*) cum1-cum52;
   EMPLOY=cumemp[week-1];
RUN;
```

The DATA step defines a new set of variables CUM1–CUM52 that are the *cumulative proportions of weeks worked* for each of the 52 weeks. The DO loop creates the cumulative number of weeks employed and converts it into a proportion. The PROC PHREG statements have the same structure as before, except that the cumulative employment (lagged by 1 week) has been substituted for the lagged employment indicator. This run produces a marginally significant effect of cumulative employment experience. When I also included the 1-week-lagged employment indicator, the effect of the cumulated variable faded to insignificance, while the lagged indicator continued to be a significant predictor. Again, it appears that the 1-week lag is a better specification.

You can create the cumulated variable in the PROC PHREG step rather than the DATA step, but that would increase computing time. In general, whatever programming *can* be done in the DATA step *should* be done there because the computations only have to be done once. In the PROC PHREG step, on the other hand, the same computations may have to be repeated many times.

Now let's repeat our analysis of the RECID data set using the counting process method. Here's the code for producing the data set:

```
DATA RECIDCOUNT;
   ARRAY emp(*) emp1-emp52;
   SET recid;
   arrest2=0;
   DO stop=1 TO week;
      start=stop-1;
      IF stop=week THEN arrest2=arrest;
      employed=emp(stop);
      IF week>1 THEN emplag=emp(stop-1); ELSE emplag=.;
      OUTPUT;
   END;
RUN;
```

This DATA step produced 19,377 person-weeks. Once again, we need an ARRAY to hold the 52 values of the employment indicator. Next, ARREST2, our new censoring indicator is initialized at 0 because individuals can only be arrested in their last person-week. The DO loop, running from week 1 until the week of arrest or censoring, creates a new person-week for each cycle through the loop. For each person-week, a START time and a STOP time are defined. For the last person-week (STOP=WEEK), the IF statement resets ARREST2 to be the same as ARREST, the original censoring indicator. Next, the variable EMPLOYED is set equal to the EMP indicator variable corresponding to that week. An employment indicator lagged by 1 week is also created. Note that here in a DATA step, we can put parentheses around the array subscript rather than the brackets that are required in the PROC PHREG step (although brackets would work also).

Because most periods of employment or unemployment extended over multiple weeks, we could have gotten by with a lot fewer records in the new data set. That is, we could have created one record for each contiguous period of employment or unemployment. But that would require a considerably more complicated program to produce the data set. In many applications, it's easier to just create a new record for each discrete point in time. This will not change the results. You *must* break the observation period whenever any covariate changes in value, but you can always break it into smaller units without affecting the outcome.

Here is the PROC PHREG step, which produces exactly the same table as the one shown in Output 5.14:

```
PROC PHREG DATA=recidcount;
MODEL (start,stop)*arrest2(0)=fin age race wexp mar paro prio employed
   /TIES=EFRON;
RUN;
```

To do the analysis with a 1-week lag of employment status, just substitute EMPLAG for EMPLOYED.

Ad-Hoc Estimates of Time-Dependent Covariates

It often happens that time-dependent covariates are measured at regular intervals, but the intervals don't correspond to the units in which *event* times are measured. For example, we may know the exact day of death for a sample of cancer patients but have only monthly measurements of, say, albumin level in the blood. For partial likelihood estimation with such data, we really need daily albumin measurements, so we must somehow impute these from the monthly data. There are often several possible ways to do this, and, unfortunately, none has any formal justification. On the other hand, it's undoubtedly better to use some common-sense method for imputing the missing values rather than discarding the data for the time-dependent covariates.

Let's consider some possible methods and some rough rules of thumb. For the case of monthly albumin measurements and day of death, an obvious method is to use the closest preceding albumin level to impute the level at any given death time. For data over a 1-year period, the SAS code for the programming statements method might look like this:

```
DATA blood;
    INFILE 'c:\blood.dat';
    INPUT deathday status alb1-alb12;
PROC PHREG;
    MODEL deathday*status(0)=albumin;
    ARRAY alb(*) alb1-alb12;
    deathmon=CEIL(deathday/30.4);
    albumin=alb[deathmon];
RUN;
```

Assume that ALB1 is measured at the *beginning* of the first month, and so on for the next 11 measurements. Dividing DEATHDAY by 30.4 converts days into months (including fractions of a month). The CEIL function then returns the smallest integer larger than its argument. Thus, day 40 would be converted to 1.32, which then becomes 2 (that is, the second month). This value is then used as a subscript in the ALB array to retrieve the albumin level recorded at the beginning of the second month. (There may be some slippage here because months vary in length. However, if we know the exact day at which each albumin measurement was taken, we can avoid this difficulty by using the methods for irregular intervals described in the next section.)

Here's how to accomplish the same thing with the counting process syntax:

```
DATA bloodcount;
  SET blood;
  ARRAY alb(*) alb1-alb12;
  status2=0;
  deathmon=CEIL(deathday/30.4);
  DO j=1 TO deathmon;
    start=(j-1)*30.4;
    stop=start+30.4;
    albumin=alb(j);
    IF j=deathmon THEN DO;
      status2=status;
      stop=deathday-start;
  END;
  OUTPUT;
END;
```

```
PROC PHREG DATA=bloodcount;
   MODEL (start,stop)*status2(0)=albumin;
RUN;
```

It may be possible to get better imputations of daily albumin levels by using information about the blood levels in earlier months. If we believe, for example, that albumin levels are likely to worsen (or improve) steadily, it might be sensible to calculate a linear extrapolation based on the most recent 2 months. Here is the programming statements code:

```
PROC PHREG DATA=blood;
   MODEL deathday*status(0)=albumin;
   ARRAY alb(*) alb1-alb12;
   deathmon=deathday/30.4;
   j=CEIL(deathmon);
   IF j=1 THEN albumin=alb(1);
   ELSE albumin=alb[j]+(alb[j]-alb[j-1])*(deathmon-j+1);
RUN;
```

To accomplish the same thing with the counting process method, the data set must consist of person months rather than person days. Here is the code:

```
DATA bloodcount;
SET blood;
   ARRAY alb(*) alb1-alb12;
   status2=0;
DO i=1 TO deathday;
   start=i-1
   stop=i;
   j=CEIL(i/30.4);
   albumin=alb(j)+(alb(j)-alb(j-1))*(i/30.4-j+1);
   IF i=deathday THEN status2=status;
   OUTPUT;
END;
PROC PHREG DATA=bloodcount;
   MODEL (start,stop)*status2(0)=albumin;
RUN;
```

In these examples, we made no use of information about albumin levels that were recorded *after* the death date. Obviously, we had no other option for patients who died. Remember, though, that for every death date, PROC PHREG retrieves (or constructs) the covariates for all individuals who were at risk of death on that date, whether or not they died. For those who did not die on that death date, we could have used the average of the albumin level recorded before the death date and the level recorded after the death date. I don't recommend this, however. Using different imputation rules for those who died and those who didn't die is just asking for artifacts to creep into your results. Even in cases where the occurrence of the event does *not* stop the measurement of the time-dependent covariate, it's a dangerous practice to use information recorded after the event to construct a variable used as a predictor of the event. This would be sensible *only* if you are completely confident that the event could not have caused any changes in the time-dependent covariate. For example, in modeling whether people will purchase a house, it might be reasonable to use an average of the local mortgage rates before and after the purchase.

Time-Dependent Covariates That Change at Irregular Intervals

In the Stanford Heart Transplant example, we had a time-dependent covariate—previous receipt of a transplant—that changed at unpredictable times. No more than one such change could occur for any of the patients. Now we consider the more general situation in which the time-dependent covariate may change at multiple, irregularly spaced points in time.

We'll do this by way of an example. The survival data (hypothetical) in Output 5.15 are for 29 males, aged 50 to 60, who were diagnosed with alcoholic cirrhosis. At diagnosis, they were measured for blood coagulation time (PT). The men were then remeasured at clinic visits that occurred at irregular intervals until they either died or the study was terminated. The maximum number of clinic visits for any patient was 10. The length of the intervals between visits ranged between 3 and 33 months, with a mean interval length of 9.5 months and a standard deviation of 6.0. In Output 5.15, SURV is the time of death or time of censoring, calculated in months since diagnosis. DEAD is coded 1 for a death and 0 if censored. TIME2–TIME10 contain the number of months since diagnosis for each clinic visit. PT1–PT10 contain the measured values of the PT variable at each clinic visit. Variables are recorded as missing if there was no clinic visit.

Output 5.15 *Survival Data for 29 Males with Alcoholic Cirrhosis*

SURV	DEAD	TIME2	TIME3	TIME4	TIME5	TIME6	TIME7	TIME8	TIME9	TIME10	PT1	PT2	PT3	PT4	PT5	PT6	PT7	PT8	PT9	PT10
90	0	7	20	26	35	44	50	56	80	83	23.9	20.8	23.6	23.6	24.0	22.5	24.6	25.1	29.4	27.9
80	0	6	36	42	54	67	78	.	.	.	29.6	15.1	15.4	16.3	13.9	14.6	16.1	.	.	.
36	0	17	28	34	.	.	.	.	.	.	25.9	24.4	24.8	24.3	.	.	.	.	.	.
68	0	15	20	26	32	51	.	.	.	.	26.8	27.9	26.5	26.5	26.8	26.6	.	.	.	.
62	0	22	40	46	.	.	.	.	.	.	23.0	25.2	27.1	27.8	.	.	.	.	.	.
47	0	5	12	24	35	46	.	.	.	.	25.8	26.0	25.2	24.9	26.3	26.6	.	.	.	.
84	0	8	27	31	43	76	.	.	.	.	14.2	11.5	12.9	12.6	12.5	18.6	.	.	.	.
57	0	6	21	27	34	39	45	51	.	.	27.6	27.5	28.0	27.8	29.1	28.2	28.3	28.4	.	.
7	0	4	7	.	.	.	.	.	.	.	25.0	25.1	24.7	.	.	.	.	.	.	.
49	0	16	.	.	.	.	.	.	.	.	25.5	27.4	.	.	.	.	.	.	.	.
55	0	3	9	15	21	33	42	49	.	.	14.8	16.7	16.9	17.7	13.8	13.8	13.7	14.2	.	.
43	0	6	11	18	24	42	.	.	.	.	20.6	19.9	20.3	20.2	19.7	27.1	.	.	.	.
42	0	6	12	18	23	29	35	.	.	.	27.6	27.0	28.1	28.8	29.0	28.4	28.8	.	.	.
11	0	9	.	.	.	.	.	.	.	.	25.3	27.8	.	.	.	.	.	.	.	.
36	1	16	22	28	.	.	.	.	.	.	22.5	22.3	25.2	26.4	.	.	.	.	.	.
36	1	9	26	32	.	.	.	.	.	.	26.9	26.9	24.2	26.2	.	.	.	.	.	.
2	1	.	.	.	.	.	.	.	.	.	19.2	.	.	.	.	.	.	.	.	.
23	1	7	13	.	.	.	.	.	.	.	21.8	20.3	23.8	.	.	.	.	.	.	.
10	1	6	.	.	.	.	.	.	.	.	21.6	22.3	.	.	.	.	.	.	.	.
29	1	21	27	.	.	.	.	.	.	.	18.7	20.2	22.5	.	.	.	.	.	.	.
16	1	6	12	.	.	.	.	.	.	.	28.4	28.7	28.7	.	.	.	.	.	.	.
15	1	7	12	.	.	.	.	.	.	.	17.8	17.7	17.4	.	.	.	.	.	.	.
5	1	3	.	.	.	.	.	.	.	.	20.7	22.6	.	.	.	.	.	.	.	.
15	1	6	.	.	.	.	.	.	.	.	28.0	28.8	.	.	.	.	.	.	.	.
1	1	.	.	.	.	.	.	.	.	.	31.6	.	.	.	.	.	.	.	.	.
13	1	4	10	.	.	.	.	.	.	.	26.0	22.7	25.4	.	.	.	.	.	.	.
39	1	22	35	.	.	.	.	.	.	.	25.5	29.0	29.2	.	.	.	.	.	.	.
20	1	12	.	.	.	.	.	.	.	.	21.3	21.0	.	.	.	.	.	.	.	.
45	1	18	24	38	.	.	.	.	.	.	23.9	28.7	29.5	30.2	.	.	.	.	.	.

Let's estimate a model in which the hazard of death at time *t* depends on the value of PT at time *t*. Because we don't have measures of PT at all death times, we'll use the closest preceding measurement. The SAS code for accomplishing this with the programming statements method is as follows:

```
PROC PHREG DATA=alco;
   MODEL surv*dead(0)=pt;
   time1=0;
   ARRAY time(*) time1-time10;
   ARRAY p(*) pt1-pt10;
   DO j=1 TO 10;
```

```
      IF surv > time[j] AND time[j] NE . THEN pt=p[j];
      END;
   RUN;
```

For a given death time, the DO loop cycles through all 10 possible clinic visits. If the death time is greater than the time of *j*th visit, the value of PT is reassigned to be the value observed at visit *j*. PROC PHREG keeps doing this until it either

- encounters a missing value of the TIME variable (no clinic visit)
- encounters a TIME value that is greater than the death time
- goes through all 10 possible visits.

Hence, PROC PHREG always stops at the most recent clinic visit and assigns the value of PT recorded at that visit.

When I ran this model, I got a coefficient for PT of .084 with a nonsignificant likelihood-ratio chi-square value of 1.93, suggesting that the blood measurement is of little predictive value. But suppose the hazard depends on the *change* in PT relative to the patient's initial measurement rather than upon the absolute level. You can implement this idea by changing the IF statement above to read as follows:

```
   IF surv > time[j] AND time[j] NE . THEN pt=p[j]-pt1;
```

With this minor change in specification, the coefficient of PT increased in magnitude to .365 with a chi-square of 7.35, which is significant at beyond the .01 level.

Here is the code for producing the same results with the counting process method:

```
   DATA alcocount;
    SET alco;
    time1=0;
    time11=.;
    ARRAY t(*) time1-time11;
    ARRAY p(*) pt1-pt10;
    dead2=0;
    DO j=1 TO 10 WHILE (t(j) NE .);
       start=t(j);
       pt=p(j);
       stop=t(j+1);
       IF t(j+1)=. THEN DO;
          stop=surv;
          dead2=dead;
       END;
       OUTPUT;
    END;
   PROC PHREG DATA=alcocount;
     MODEL (start,stop)*dead2(0)=pt;
   RUN;
```

To make the predictor measure the change from TIME1, simply modify the definition of PT to be

```
pt=p(j)-pt1;
```

In general, the only additional requirement for handling covariates with irregular intervals between measurements is data on the times of measurement. The variables containing the measurement times must always be in the same metric as the event-time variable.

COX MODELS WITH NONPROPORTIONAL HAZARDS

Earlier, I mentioned that the Cox model can be easily extended to allow for nonproportional hazards. In fact, we've just finished a lengthy discussion of one class of nonproportional models. Whenever you introduce time-dependent covariates into a Cox regression model, it's no longer accurate to call it a proportional hazards (PH) model. Why? Because the time-dependent covariates will change at different rates for different individuals, so the ratios of their hazards cannot remain constant. As we've seen, however, that creates no real problem for the partial likelihood estimation method.

But suppose you don't have any time-dependent covariates. How do you know whether your data satisfy the PH assumption, and what happens if the assumption is violated? Although these are legitimate questions, I personally believe that concern about the PH assumption is often excessive. Every model embodies many assumptions, some more questionable or consequential than others. The reason people focus so much attention on the PH assumption is that the model is *named* for that property. At the same time, they often ignore such critical questions as: Are all the relevant covariates included? Is the censoring mechanism noninformative? Is measurement error in the covariates acceptably low? As in ordinary linear regression, measurement error in the covariates tends to attenuate coefficients (Nakamura, 1992).

To put this issue in perspective, you need to understand that violations of the PH assumption are equivalent to interactions between one or more covariates and time. That is, the PH model assumes that the effect of each covariate is the same at all points in time. If the effect of a variable varies with time, the PH assumption is violated for that variable. It's unlikely that the PH assumption is ever exactly satisfied, but that's true of nearly all statistical assumptions. If we estimate a PH model when the assumption is violated for some variable (thereby suppressing the interaction), then the coefficient that we estimate for that variable is a sort of *average* effect over the range of times observed in the data. Is this so terrible? In fact, researchers suppress interactions all the time when they estimate regression models. In models with even a moderately large number of variables, no one tests for all the possible 2-way interactions—there are just too many of them.

In some cases, however, we may have reason to believe that the interactions with time are so strong that it would be misleading to suppress them. In other situations, we might have a strong theoretical interest in those interactions, even when they are not very strong. For those cases, we need to examine methods for testing and modeling nonproportional hazards.

We begin with methods for testing for nonproportional hazards using residuals. Then we examine two methods that extend the Cox model to allow for nonproportional hazards. One method explicitly incorporates the interactions into the model. The other, called *stratification*, subsumes the interactions into the arbitrary function of time.

Testing the Proportionality Assumption with the ASSESS Statement

A variety of different kinds of residuals have been defined for Cox regression models (Collett, 2003). One kind, *martingale residuals*, can be used to test for nonproportionality, using the method proposed by Lin, Wei, and Ying (1993). This method has been incorporated into the ASSESS statement in PROC PHREG. Unfortunately, this statement will not work if the model contains time-dependent covariates, either with the programming statements method or the counting process method.

Here's how to use the ASSESS statement for the recidivism data:

```
ODS GRAPHICS ON;
PROC PHREG DATA=recid;
   MODEL week*arrest(0)=fin age race wexp mar paro prio
         / TIES=EFRON;
   ASSESS PH / RESAMPLE;
RUN;
ODS GRAPHICS OFF;
```

In the ASSESS statement, the PH option requests an assessment of the proportional hazards assumption. For each covariate, this statement produces a graphical display of the *empirical score process*, which is based on the martingale residuals. Output 5.16 displays the graph for AGE. The solid line is the observed empirical score process. The dashed lines are empirical score processes based on 20 random simulations that embody the proportional hazards assumption. If the observed process deviates markedly from the simulated processes, it is evidence against the proportional hazards assumption. The observed process for AGE does indeed appear to be somewhat more extreme than the simulated processes.

Output 5.16   *Testing the Proportional Hazards Assumption with the ASSESS Statement*

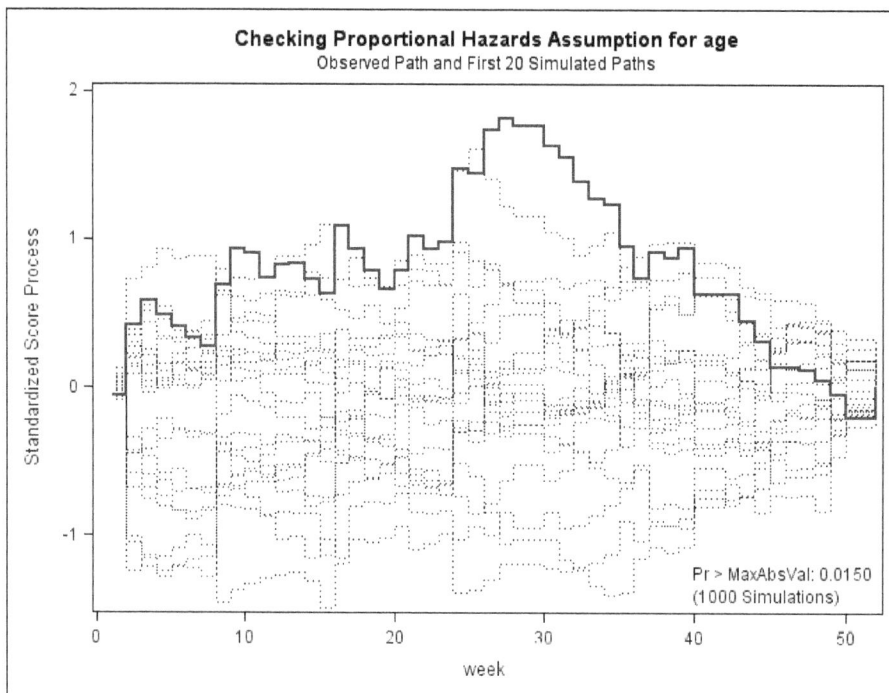

In the lower right corner of Output 5.16, we get a more quantitative assessment in the form of a *p*-value. For 1000 simulated paths, only 1.5 percent of them had extreme points (suprema) that exceeded the most extreme point of the observed path (yielding a Kolmogorov-Smirnov type test). The *p*-value was produced by the RESAMPLE option, which generated the 1,000 simulated paths. That option also produces a table summarizing the results for each covariate, as seen in Output 5.17. The only variable with clear evidence for a violation of the PH assumption is AGE. However, this test does not tell us anything about the nature of the violation. Note also that if you only want the table but not the graphs, you can just run the code above without the ODS statements.

Output 5.17   *Testing the Proportional Hazards Assumption with the ASSESS Statement*

```
              Supremum Test for Proportional Hazards Assumption

                      Maximum
                      Absolute                                    Pr >
         Variable       Value    Replications      Seed        MaxAbsVal

           fin         0.5408        1000        292778001       0.8240
           age         1.8192        1000        292778001       0.0150
           race        0.9435        1000        292778001       0.2210
           wexp        1.3008        1000        292778001       0.0940
           mar         0.9349        1000        292778001       0.2350
           paro        0.5383        1000        292778001       0.8070
           prio        0.6104        1000        292778001       0.7240
```

You can also test for nonproportional hazards using Schoenfeld residuals. Unlike the ASSESS statement, this method works even if you have time-varying covariates defined by the counting process method (but not with the programming statements method). It also takes much less computing time, which may be important with large samples.

Schoenfeld residuals have an unusual property: instead of a single residual for each individual, there is a separate residual for each *covariate* for each individual. They are not defined for censored observations (they are missing in the output data set). Here's how they are calculated. Suppose that individual i is arrested at time t_i, and at that time there were 330 people at risk of arrest, indexed by $j = 1, \ldots, 330$. For each of those 330 people, the estimated Cox model implies a certain probability of being arrested at that time, denoted by p_j. Imagine randomly selecting one of these 330 people, with probability p_j. For each covariate x_k, we can calculate its expected value for a randomly selected person (from that risk set) as

$$\bar{x}_k = \sum_{j=1}^{330} x_{kj} p_j \; .$$

The Schoenfeld residual is then defined as the covariate value for the person who *actually* died, x_{ik}, minus the expected value.

The main use of these residuals is to detect possible departures from the proportional hazards assumption. If the assumption is satisfied, the Schoenfeld residuals should be independent of time. Testing the proportional hazards assumption is equivalent, therefore, to testing whether the Schoenfeld residuals are correlated with time or with some function of time. For the recidivism data, I produced a data set containing the Schoenfeld residuals using these statements:

```
PROC PHREG DATA=recid;
  MODEL week*arrest(0)=fin age prio race wexp mar paro prio /
    TIES=EFRON;
OUTPUT OUT=b RESSCH=schfin schage schprio schrace schwexp schmar
  schparo schprio;
RUN;
```

Note that the keyword RESSCH is followed by eight arbitrarily chosen variable names corresponding to the eight covariates in the model. These variables contain the Schoenfeld residuals in the output data set B.

The next step is to modify the data set to incorporate two familiar functions of time, the log and the square, followed by PROC CORR to compute the correlations of the residuals with time and its two functions. Other functions of time might also be considered.

```
DATA c;
  SET b;
  lweek=LOG(week);
  week2=week**2;
PROC CORR;
VAR week lweek week2 schfin schage schprio schrace schwexp schmar
  schparo schprio;
RUN;
```

The results, shown in Output 5.18, are consistent with what we found using the ASSESS statement. There is a fairly clear-cut problem with AGE (with *p*-values less than .05 for all three functions) and a marginal problem with WEXP. Notice that the sample size is only 114 because the Schoenfeld residuals are not defined for the censored cases. In the next subsection, we will see how to estimate a Cox model that allows the effect of AGE to change with time. This will also give us a more definitive test of the proportionality assumption for this variable.

Output 5.18 *Testing the Proportional Hazards Assumption with Schoenfeld Residuals*

```
                     Pearson Correlation Coefficients
                       Prob > |r| under H0: Rho=0
                          Number of Observations

                              week       1week       week2

  schfin                    -0.00874     0.03484    -0.01688
  Schoenfeld Residual for fin  0.9265     0.7129      0.8585
                                 114         114         114

  schage                    -0.21563    -0.23849    -0.18679
  Schoenfeld Residual for age  0.0212     0.0106      0.0466
                                 114         114         114

  schprio                   -0.06043     0.01215    -0.08285
  Schoenfeld Residual for prio 0.5230     0.8979      0.3809
                                 114         114         114

  schrace                   -0.13550    -0.11724    -0.13527
  Schoenfeld Residual for race 0.1506     0.2141      0.1513
                                 114         114         114

  schwexp                    0.18432     0.13489     0.20662
  Schoenfeld Residual for wexp 0.0496     0.1524      0.0274
                                 114         114         114

  schmar                     0.09382     0.09654     0.08061
  Schoenfeld Residual for mar  0.3208     0.3069      0.3939
                                 114         114         114

  schparo                    0.01596     0.04421     0.00346
  Schoenfeld Residual for paro 0.8661     0.6404      0.9709
                                 114         114         114

  schprio                   -0.06043     0.01215    -0.08285
  Schoenfeld Residual for prio 0.5230     0.8979      0.3809
                                 114         114         114
```

INTERACTIONS WITH TIME AS TIME-DEPENDENT COVARIATES

A common way of representing interaction between two variables in a linear regression model is to include a new variable that is the product of the two variables in question. To represent the interaction between a covariate x and time in a Cox model, we can write

$$\log h(t) = \alpha(t) + \beta_1 x + \beta_2 xt .$$

Factoring out the x, we can rewrite this as

$$\log h(t) = \alpha(t) + (\beta_1 + \beta_2 t) x .$$

In this equation, the effect of x is $\beta_1 + \beta_2 t$. If β_2 is positive, then the effect of x increases linearly with time; if it's negative, the effect decreases linearly with time. β_1 can be interpreted as the effect of x at time 0, the origin of the process.

You can easily estimate this model by defining a time-dependent covariate $z=xt$. Here's an example. Using both the ASSESS statement and Schoenfeld residuals, we just saw evidence that the variable AGE violates the proportional hazards assumption, implying that the effect of AGE varies with time since release. To incorporate that interaction using the programming statements method, we can run the following SAS code, which yields the results in Output 5.19:

```
PROC PHREG DATA=recid;
   MODEL week*arrest(0)=fin age race wexp mar paro prio ageweek
         / TIES=EFRON;
   ageweek=age*week;
RUN;
```

Output 5.19 *A Model for Recidivism with Time by Age Interaction*

Parameter	DF	Parameter Estimate	Standard Error	Chi-Square	Pr > ChiSq	Hazard Ratio
fin	1	-0.37823	0.19129	3.9096	0.0480	0.685
age	1	0.03692	0.03917	0.8883	0.3459	1.038
race	1	0.32290	0.30804	1.0989	0.2945	1.381
wexp	1	-0.12224	0.21285	0.3298	0.5658	0.885
mar	1	-0.41162	0.38212	1.1604	0.2814	0.663
paro	1	-0.09293	0.19583	0.2252	0.6351	0.911
prio	1	0.09354	0.02869	10.6294	0.0011	1.098
ageweek	1	-0.00369	0.00146	6.4241	0.0113	0.996

The interaction term is significant at about the .01 level, which confirms what we found using the ASSESS statement and the Schoenfeld residuals. But now we can go further and say something specific about how the effect of AGE varies with time. The main effect of AGE is the effect of that variable when WEEK is 0 (that is, when people are first released from prison). The hazard ratio tells us that each additional year of AGE increases the hazard of arrest by about 3.8 percent, although that number is not significantly different from 0. To get the effect of age for any given week, we can calculate .03692 − .00369*WEEK. For example, at 10 weeks from release, the effect is approximately 0. At 20 weeks, it's −.0369. At 30 weeks, it's −.0738. And at 40 weeks, it's −.1107. We can test whether the effect of AGE is significantly different from 0 at any given week by

using a TEST statement. For example, to test the effect at week 30, the
TEST statement would be

```
TEST age+30*ageweek=0;
```

This returns a chi-square value of 9.06, with a *p*-value of .0026.

Some textbooks recommend forming the product of the covariate
and the *logarithm* of time, possibly to avoid numerical overflows. But
there is no theoretical reason to prefer this specification, and the algorithm
used by PROC PHREG is robust to numerical problems. When I used the
logarithm of time with the recidivism data, the chi-square for the
interaction was higher but not by much.

As an alternative to treating time as continuous in the interaction, I
defined it as dichotomous with the following programming statement:

```
ageweek=age*(week>26);
```

In this case, the main effect of AGE (–.004 with a *p*-value of .88) may be
interpreted as the effect of AGE during the first 6 months of the year. To get
the effect of AGE during the second 6x months, we add the interaction to
the main effect, which yields a value of –.145 with a *p*-value of .0004. So it
appears that AGE has essentially no effect on the hazard of arrest during
the first 6 months but a big negative effect during the second 6 months.

To sum up, we now have a definitive way to test for violations of the
PH assumption: For any suspected covariate, simply add to the model a
time-dependent covariate representing the interaction of the original
covariate and time. If the interaction covariate does not have a significant
coefficient, then we may conclude that the PH assumption is not violated
for that variable. On the other hand, if the interaction variable does have a
significant coefficient, we have evidence for nonproportionality. Of course,
we also have a model that incorporates the nonproportionality. So, in this
case, the method of diagnosis is also the cure.

NONPROPORTIONALITY VIA STRATIFICATION

Another approach to nonproportionality is *stratification*, a technique
that is most useful when the covariate that interacts with time is both
categorical and not of direct interest. Consider the myelomatosis data
described in Chapter 2 and analyzed in Chapter 3. For that data, the
primary interest was in the effect of the treatment indicator TREAT. There
was also a covariate called RENAL, an indicator of whether renal
functioning was impaired at the time of randomization to treatment. In

Chapter 3, we found that RENAL was strongly related to survival time, but the effect of TREAT was not significant. Suppose we believe (or suspect) that the effect of RENAL varies with time since randomization. Alternatively, we can say that the shape of the hazard function is different for those with and without impaired renal functioning. Letting x represent the treatment indicator and z the renal functioning indicator, we can represent this idea by postulating separate models for the two renal functioning groups:

<u>Impaired</u>   <u>Not Impaired</u>
$$\log h_i(t) = \alpha_0(t) + \beta x_i \qquad\qquad \log h_i(t) = \alpha_1(t) + \beta x_i$$

Notice that the coefficient of x is the same in both equations, but the arbitrary function of time is allowed to differ. We can combine the two equations into a single equation by writing

$$\log h_i(t) = \alpha_z(t) + \beta x_i .$$

One can easily estimate this model by the method of partial likelihood using these steps:

1. Construct separate partial likelihood functions for each of the two renal functioning groups.
2. Multiply those two functions together.
3. Choose values of β that maximize this function.

PROC PHREG does this automatically if you include a STRATA statement:

```
PROC PHREG DATA=myel;
   MODEL dur*stat(0)=treat / TIES=EXACT;
   STRATA renal;
RUN;
```

Output 5.20 shows the results. PROC PHREG first reports some simple statistics for each of the strata; then it reports the usual output for the regression model. In contrast to the PROC LIFETEST results of Chapter 3, we now find a statistically significant effect of the treatment. Notice, however, that there is no estimated coefficient for RENAL. Whatever effects this variable may have, they are entirely absorbed into the two arbitrary functions of time.

Output 5.20 *Myelomatosis Data with Stratified Cox Regression*

```
                Summary of the Number of Event and Censored Values

                                                              Percent
       Stratum   renal        Total      Event    Censored    Censored

          1      0              18         10          8        44.44
          2      1               7          7          0         0.00
       ------------------------------------------------------------------
       Total                    25         17          8        32.00




                    Testing Global Null Hypothesis: BETA=0

           Test                  Chi-Square      DF     Pr > ChiSq

           Likelihood Ratio        6.0736         1        0.0137
           Score                   5.7908         1        0.0161
           Wald                    4.9254         1        0.0265


                    Analysis of Maximum Likelihood Estimates

                        Parameter   Standard                             Hazard
       Parameter   DF    Estimate      Error   Chi-Square   Pr > ChiSq    Ratio

       treat        1     1.46398    0.65965      4.9254       0.0265      4.323
```

For comparison purposes, I also estimated three other PROC PHREG models:

- a model with TREAT as the only covariate and no stratification. The coefficient was .56, with a *p*-value (based on the Wald chi-square test) of .27, which is approximately the same as the *p*-value for the log-rank test reported in Chapter 3.
- a model with both TREAT and RENAL as covariates. The coefficient of TREAT was 1.22, with a *p*-value of .04.
- a model that included TREAT, RENAL, and the (time-varying) product of RENAL and DUR. The coefficient of TREAT was 1.34, with a *p*-value of .05. The interaction term was not significant.

Based on these results, it is clearly essential to control for RENAL in order to obtain good estimates of the treatment effect, but whether you do it by stratification or by including RENAL as a covariate doesn't seem to make much difference. Note that you cannot stratify by a variable and also include it as a covariate. These are alternative ways of controlling for the variable.

Compared with the explicit interaction method of the previous section, the method of stratification has two main advantages:

- The interaction method requires that you choose a particular form for the interaction, but stratification allows for any change in the effect of a variable over time. For example, including the product of RENAL and DUR as a covariate forces the effect of RENAL to either increase linearly or decrease linearly with time. Stratification by RENAL allows for reversals (possibly more than one) in the relationship between time and the effect of RENAL.

- Stratification takes less computing time. This can be important in working with large samples.

But there are also important disadvantages of stratification:

- No estimates are obtained for the effect of the stratifying variable. As a result, stratification only makes sense for nuisance variables whose effects have little or no intrinsic interest.

- There is no way to test for either the main effect of the stratifying variable or its interaction with time. In particular, it is *not* legitimate to compare the log-likelihoods for models with and without a stratifying variable.

- If the form of the interaction with time is correctly specified, the explicit interaction method should yield more efficient estimates of the coefficients of the other covariates. Again, there is a trade-off between robustness and efficiency.

Now for some complications. In the example we've just been considering, the stratifying variable had only two values. If a variable with more than two values is listed in the STRATA statement, PROC PHREG creates a separate stratum for each value. The stratifying variable can be either character or numeric, with no restrictions on the possible values. If more than one variable is listed in the STRATA statement, PROC PHREG creates a separate stratum for every observed combination of values. Obviously you must exercise some care in using this option because you can easily end up with strata that have only one observation. For stratifying variables that are numeric, it is also easy to specify a limited number of strata that correspond to cut points on the variable. The syntax for accomplishing this is identical to the description in Chapter 3 for the STRATA statement in PROC LIFETEST.

In my view, the most useful application of stratification is for samples involving some kind of clustering or multi-level grouping. Examples include animals within litters, children within schools, patients within hospitals, and so on. In all of these situations, it is reasonable to expect that observations within a cluster will not be truly independent, thereby violating one of the standard assumptions used in constructing the likelihood function. The result is likely to be standard error estimates that are biased downward and test statistics that are biased upward. A highly effective solution is to treat each cluster as a distinct stratum. You can accomplish this by giving every cluster a unique identification number and then stratifying on the variable containing that number. This method assumes that observations are *conditionally* independent within clusters and that the coefficients of the covariates are the same across clusters. (The conditional independence assumption would be violated if, for example, the death of one patient had a deleterious effect on the risk of death for other patients in the same hospital.) In Chapter 8, we will see an alternative method for dealing with clustering.

LEFT TRUNCATION AND LATE ENTRY INTO THE RISK SET

In standard treatments of partial likelihood, it is assumed that every individual is at risk of an event at time 0 and continues to be at risk until either the event of interest occurs or the observation is censored. At that point the individual is removed from the risk set, never to return. It doesn't have to be this way, however. If there are periods of time when it is known that a particular individual is not at risk for some reason, the partial likelihood method allows for the individual to be removed from the risk set and then returned at some later point in time. Unfortunately, many partial likelihood programs are not set up to handle this possibility.

The most common situation is *left truncation* or *late entry to the risk set*. This situation often arises without the investigator even realizing it. Here are three examples:

- Patients are recruited into a study at varying lengths of time after diagnosis with some disease. The investigator wishes to specify a model in which the hazard of death depends on the time since diagnosis. But by the design of the study, it is impossible for the patients to die between diagnosis and recruitment. If they had died, they would not have been available for recruitment.

- At a specific point in time, all employees in a firm are interviewed and asked how long they have been working for the firm. Then, they are followed forward to determine how long they stay with the firm. The investigator estimates a model in which the hazard of termination depends on the length of employment with the firm. But again, it is impossible for these employees to have terminated before the initial interview. If they had terminated, they would not have been present for that interview.

■ High school girls are recruited into a study of teen pregnancy. The desired model expresses the hazard of pregnancy as a function of age. But girls were not included in the study if they had become pregnant before the study began. Thus, the girls in the study were not *observationally* at risk of pregnancy at ages before the initial interview. Obviously, they were at risk during those ages, but the design of the study makes it impossible to observe pregnancies that occurred before the initial interview.

In all three cases, the solution is to remove the individual from the risk set between the point of origin and the time of the initial contact, which you can easily accomplish using the counting process syntax in PROC PHREG.

To illustrate these methods, let's specify a model for the Stanford Heart Transplant Data in which the hazard is expressed as a function of patient age rather than time since acceptance into the program. To do this, we must first convert the death time variable into patient age (in years):

```
DATA stan2;
   SET stan;
   agels=(dls-dob)/365.25;
RUN;
```

Recall that DLS is date last seen and DOB is date of birth. Then AGELS is age at which the patient was last seen, which may be either a death time or a censoring time. Our problem is that, by design, patients were not at risk of an observable death before they were accepted into the program.

Using the counting process syntax, we can specify a model in the following way:

```
PROC PHREG DATA=stan2;
   MODEL (ageaccpt,agels)*dead(0)=surg ageaccpt / TIES=EFRON;
RUN;
```

Here, AGEACCPT is the time of entry into the risk set and AGELS is the time of departure. Note that AGEACCPT can also be included as a covariate.

It is also possible to combine the counting process syntax with the programming statements method for defining a time-dependent covariate.

The following program defines PLANT as a time-dependent indicator of transplant status, with results shown in Output 5.21:

```
PROC PHREG DATA=stan2;
   MODEL (ageaccpt,agels)*dead(0)=plant surg ageaccpt / TIES=EFRON;
   IF agetrans>=agels OR agetrans=. THEN plant=0;
   ELSE plant=1;
RUN;
```

Compare the results in Output 5.21 with those in Output 5.11, which had the same covariates but specified the principal time axis as time since acceptance. For PLANT and SURG, the results are very similar. But there is a dramatic increase in the effect of age at acceptance, which is a direct consequence of specifying age as the time axis. With age held constant (every risk set consists of people who are all the same age), age at acceptance is an exact linear function of *time* since acceptance. In the earlier analysis, any effects of time since acceptance were part of the baseline hazard function and were, therefore, not visible in the estimated coefficients.

Output 5.21 *Results for Transplant Data with Age as the Principal Time Axis*

Parameter	DF	Parameter Estimate	Standard Error	Chi-Square	Pr > ChiSq	Hazard Ratio
plant	1	-0.28042	0.43761	0.4106	0.5217	0.755
surg	1	-0.85312	0.46529	3.3618	0.0667	0.426
ageaccpt	1	0.75857	0.28338	7.1655	0.0074	2.135

So far, I have explained how to handle situations in which individuals who are *not* in the risk set at time 0 enter the risk set at some later point in time and then remain in the risk set until an event occurs. Now let's consider situations in which individuals temporarily leave the risk set for some reason and then return at a later point in time. Suppose, for example, that a sample of children is followed for a period of 5 years to detect any occurrences of parental abuse. If some children are placed in foster homes for temporary periods, it may be desirable to remove them from the risk set for those periods.

To accomplish this, a separate record is created for each interval during which the individual was continuously at risk. Thus, for a child who had one temporary period in a foster home, the first record corresponds to the interval from time 0 (however defined) to the time of entry into the foster home. This record is coded as censored because the event of interest did not occur. The second record is for the interval from the time of exit from the foster home to the last time of observation, coded as censored if no event occurred at the termination of observation. The MODEL statement then uses the counting process syntax:

```
MODEL (start, finish)*CENSOR(0)=X1 X2;
```

Here, *start* is the time at which the interval begins (measured from the origin) and *finish* is the time at which the interval ends. The extension to multiple time outs should be straightforward.

ESTIMATING SURVIVOR FUNCTIONS

As we have seen, the form of the dependence of the hazard on time is left unspecified in the proportional hazards model. Furthermore, the partial likelihood method discards that portion of the likelihood function that contains information about the dependence of the hazard on time. Nevertheless, it is still possible to get nonparametric estimates of the survivor function based on a fitted proportional hazards model.

When there are no time-dependent covariates, the Cox model can be written as

$$S_i(t) = \left[S_0(t) \right]^{\exp(\beta \mathbf{x}_i)} \tag{5.14}$$

where $S_i(t)$ is the probability that individual i with covariate values $\mathbf{x}_i$ survives to time t. $S_0(t)$ is called the *baseline survivor function* (that is, the survivor function for an individual whose covariate values are all 0). After β is estimated by partial likelihood, $S_0(t)$ can be estimated by a nonparametric maximum likelihood method. With that estimate in hand, you can generate the estimated survivor function for any set of covariate values by substitution into equation (5.14).

In PROC PHREG, this is accomplished with the BASELINE statement and the PLOTS option. The easiest task is to get the survivor function for $\mathbf{x}_i = \bar{\mathbf{x}}$, the vector of sample means. For the recidivism data set, the SAS code for accomplishing this task is as follows:

```
ODS GRAPHICS ON;
PROC PHREG DATA=recid PLOTS=S;
   MODEL week*arrest(0)=fin age prio
        / TIES=EFRON;
   BASELINE OUT=a SURVIVAL=s LOWER=lcl UPPER=ucl;
RUN;
ODS GRAPHICS OFF;
```

In the BASELINE statement, the OUT=A option puts the survival estimates in a temporary data set named A. SURVIVAL=S stores the survival probabilities in a variable named S. The UPPER and LOWER options store the upper and lower 95% confidence limits in variables UCL and LCL, respectively.

Data set A is printed in Output 5.22. There are 50 records, corresponding to the 49 unique weeks in which arrests were observed to occur, plus an initial record for time 0. Each record gives the mean value for each of the three covariates. The S column gives the estimated survival probabilities. The last two columns are the upper and lower 95% confidence limits. The survivor function reported here is very similar to one that would be produced by PROC LIFETEST using the Kaplan-Meier method, but it is not identical. This one allows for heterogeneity in the hazard/survivor functions while the Kaplan-Meier estimator does not.

Output 5.22 *Survivor Function Estimates for Recidivism Data at Sample Means*

Obs	fin	age	prio	week	s	lcl	ucl
1	0.5	24.5972	2.98380	0	1.00000	.	.
2	0.5	24.5972	2.98380	1	0.99801	0.99411	1.00000
3	0.5	24.5972	2.98380	2	0.99602	0.99050	1.00000
4	0.5	24.5972	2.98380	3	0.99403	0.98728	1.00000
5	0.5	24.5972	2.98380	4	0.99204	0.98425	0.99989
6	0.5	24.5972	2.98380	5	0.99005	0.98134	0.99884
7	0.5	24.5972	2.98380	6	0.98805	0.97851	0.99769
8	0.5	24.5972	2.98380	7	0.98605	0.97573	0.99648
9	0.5	24.5972	2.98380	8	0.97610	0.96258	0.98981
10	0.5	24.5972	2.98380	9	0.97209	0.95747	0.98694
11	0.5	24.5972	2.98380	10	0.97009	0.95494	0.98547
12	0.5	24.5972	2.98380	11	0.96606	0.94991	0.98248
13	0.5	24.5972	2.98380	12	0.96197	0.94487	0.97939
14	0.5	24.5972	2.98380	13	0.95993	0.94237	0.97783
15	0.5	24.5972	2.98380	14	0.95381	0.93494	0.97307
16	0.5	24.5972	2.98380	15	0.94970	0.92999	0.96983
17	0.5	24.5972	2.98380	16	0.94558	0.92507	0.96654
18	0.5	24.5972	2.98380	17	0.93942	0.91778	0.96158
19	0.5	24.5972	2.98380	18	0.93321	0.91047	0.95651
20	0.5	24.5972	2.98380	19	0.92904	0.90560	0.95309
21	0.5	24.5972	2.98380	20	0.91865	0.89356	0.94445
22	0.5	24.5972	2.98380	21	0.91448	0.88876	0.94096
23	0.5	24.5972	2.98380	22	0.91240	0.88636	0.93920
24	0.5	24.5972	2.98380	23	0.91031	0.88396	0.93744
25	0.5	24.5972	2.98380	24	0.90198	0.87445	0.93037

(*continued*)

Output 5.22 *(continued)*

26	0.5	24.5972	2.98380	25	0.89573	0.86735	0.92504
27	0.5	24.5972	2.98380	26	0.88938	0.86017	0.91959
28	0.5	24.5972	2.98380	27	0.88514	0.85539	0.91593
29	0.5	24.5972	2.98380	28	0.88090	0.85062	0.91227
30	0.5	24.5972	2.98380	30	0.87665	0.84584	0.90857
31	0.5	24.5972	2.98380	31	0.87451	0.84345	0.90672
32	0.5	24.5972	2.98380	32	0.87024	0.83867	0.90300
33	0.5	24.5972	2.98380	33	0.86596	0.83389	0.89925
34	0.5	24.5972	2.98380	34	0.86165	0.82909	0.89547
35	0.5	24.5972	2.98380	35	0.85302	0.81953	0.88788
36	0.5	24.5972	2.98380	36	0.84653	0.81235	0.88214
37	0.5	24.5972	2.98380	37	0.83787	0.80281	0.87446
38	0.5	24.5972	2.98380	38	0.83569	0.80042	0.87252
39	0.5	24.5972	2.98380	39	0.83133	0.79563	0.86863
40	0.5	24.5972	2.98380	40	0.82262	0.78610	0.86085
41	0.5	24.5972	2.98380	42	0.81823	0.78130	0.85691
42	0.5	24.5972	2.98380	43	0.80948	0.77177	0.84903
43	0.5	24.5972	2.98380	44	0.80505	0.76696	0.84503
44	0.5	24.5972	2.98380	45	0.80061	0.76214	0.84102
45	0.5	24.5972	2.98380	46	0.79171	0.75251	0.83294
46	0.5	24.5972	2.98380	47	0.78947	0.75010	0.83091
47	0.5	24.5972	2.98380	48	0.78500	0.74528	0.82684
48	0.5	24.5972	2.98380	49	0.77384	0.73328	0.81665
49	0.5	24.5972	2.98380	50	0.76712	0.72608	0.81049
50	0.5	24.5972	2.98380	52	0.75812	0.71644	0.80221

The PLOTS=S option produces a graph of the adjusted survivor function, evaluated at the means of the covariates, displayed in Output 5.23. The PLOTS option only works with ODS graphics, which is why the PROC PHREG step is preceded and followed by ODS statements. The "Reference Setting" in the graph title refers to the mean values of the covariates. However, if the MODEL statement contains any CLASS variables, the survivor function is evaluated at the reference levels of those variables rather than at their means. In a moment, we will see how to evaluate and graph the survivor function at other specified covariate values.

Output 5.23 *Graph of the Adjusted Survivor Function for Recidivism Data*

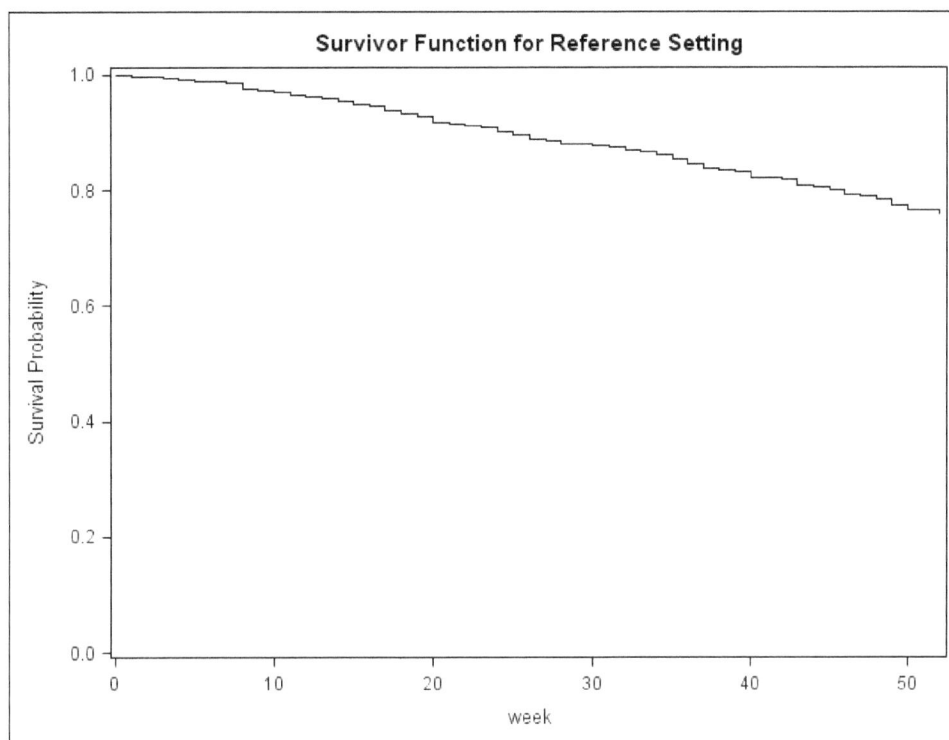

By combining the BASELINE statement with stratification, we can also produce graphs that compare survivor functions for two or more groups, while adjusting for covariates. Suppose we take financial aid (FIN) as the stratifying variable for the recidivism data. That might seem self-defeating because FIN is the variable of greatest interest and stratifying on it implies that no tests or estimates of its effect are produced. But after stratifying, we can graph the baseline survivor functions for the two financial aid groups using the following code:

```
ODS GRAPHICS ON;
PROC PHREG DATA=recid PLOTS(OVERLAY=ROW)=S;
   MODEL week*arrest(0)=fin age prio
        / TIES=EFRON;
   STRATA fin;
RUN;
ODS GRAPHICS OFF;
```

The OVERLAY option is necessary to get the two survival curves on the same graph. Note that the BASELINE statement used in the previous program is not necessary if you only want the graphs. The resulting figure in Output 5.24 shows the survivor functions for each of the two financial aid groups, evaluated at the means of the covariates.

Output 5.24 *Survivor Plots for the Two Financial Aid Groups*

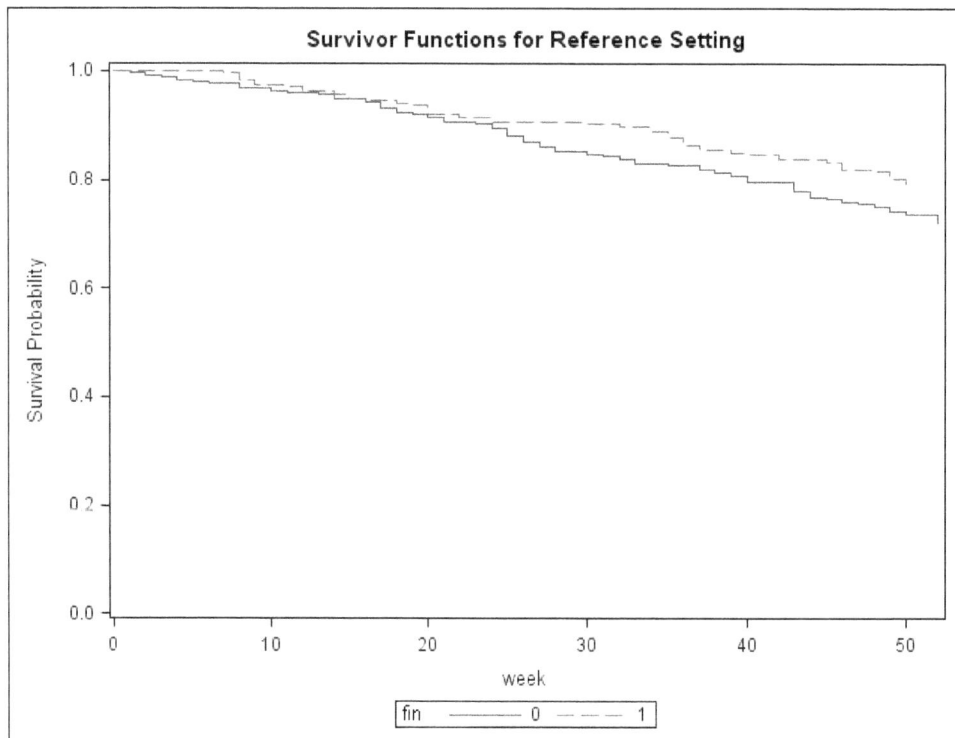

Another major use of the baseline survivor function is to obtain predictions about survival time for particular sets of covariate values. These covariate values need not appear in the data set being analyzed. For the recidivism data, for example, we may want to say something about arrest times for 20-year-olds with four prior convictions who did not receive financial aid. The mechanics of doing this are a bit awkward. You must first create a new data set containing the values of the covariates for which you want predictions and then pass the name of that data set to PROC PHREG:

```
DATA covals;
   INPUT fin age prio;
   DATALINES;
0 40 3
;
```

```
PROC PHREG DATA=recid;
   MODEL week*arrest(0)=fin age prio / TIES=EFRON;
   BASELINE OUT=a COVARIATES=covals SURVIVAL=s LOWER=lcl
        UPPER=ucl;
PROC PRINT DATA=a;
RUN;
```

The advantage of doing it this way is that predictions can easily be generated for many different sets of covariate values just by including more input lines in the data set COVALS. Each input line produces a complete set of survivor estimates, but all estimates are output to a single data set. The LOWER= and UPPER= options give 95% confidence intervals around the survival probability.

Output 5.25 displays a portion of the data set generated by the BASELINE statement above. In generating predictions, it's typical to focus on a single summary measure rather than the entire distribution. The median survival time is easily obtained by finding the smallest value of t such that $S(t) \leq .50$. That won't work for the recidivism data, however, because the data are censored long before a .50 probability is reached. For these data, it's probably more useful to pick a fixed point in time and calculate survival probabilities at that time under varying conditions. For the covariate values in Output 5.25, the 6-month (26-week) survival probability is .82, with a 95% confidence interval of .77 to .88.

Output 5.25   *Portion of Survivor Function Estimate for Recidivism Data*

Obs	fin	age	prio	week	s	lcl	ucl
1	0	20	4	0	1.00000	.	.
2	0	20	4	1	0.99644	0.98948	1.00000
3	0	20	4	2	0.99289	0.98305	1.00000
4	0	20	4	3	0.98935	0.97730	1.00000
5	0	20	4	4	0.98582	0.97189	0.99994
6	0	20	4	5	0.98229	0.96671	0.99811
7	0	20	4	6	0.97875	0.96168	0.99613
8	0	20	4	7	0.97522	0.95675	0.99404
9	0	20	4	8	0.95769	0.93345	0.98257
10	0	20	4	9	0.95068	0.92444	0.97767
11	0	20	4	10	0.94718	0.92000	0.97517
12	0	20	4	11	0.94017	0.91117	0.97009
13	0	20	4	12	0.93308	0.90232	0.96489
14	0	20	4	13	0.92954	0.89794	0.96226
15	0	20	4	14	0.91898	0.88497	0.95429
16	0	20	4	15	0.91191	0.87636	0.94890

(continued)

Output 5.25 *(continued)*

17	0	20	4	16	0.90486	0.86783	0.94346
18	0	20	4	17	0.89435	0.85524	0.93525
19	0	20	4	18	0.88381	0.84267	0.92695
20	0	20	4	19	0.87677	0.83433	0.92137
21	0	20	4	20	0.85932	0.81381	0.90738
22	0	20	4	21	0.85237	0.80570	0.90175
23	0	20	4	22	0.84891	0.80166	0.89894
24	0	20	4	23	0.84543	0.79762	0.89611
25	0	20	4	24	0.83166	0.78169	0.88482
26	0	20	4	25	0.82140	0.76989	0.87634
27	0	20	4	26	0.81102	0.75799	0.86776
28	0	20	4	27	0.80413	0.75011	0.86204
29	0	20	4	28	0.79726	0.74227	0.85631
30	0	20	4	30	0.79039	0.73445	0.85058

Note that the BASELINE statement will not produce any output when there are time-dependent covariates defined by the programming statements method. The BASELINE statement *will* produce output when time-dependent covariates are handled with the counting process method, although the log window will contain the following warning: `NOTE: Since the counting process style of response was specified in the MODEL statement, the SURVIVAL= statistics in the BASELINE statement should be used with caution.` For most applications, however, this should not be a concern.

TESTING LINEAR HYPOTHESES WITH CONTRAST OR TEST STATEMENTS

In Chapter 4, I discussed the CLASS statement in PROC LIFEREG, and I showed how to do non-standard hypothesis tests using the estimated covariance matrix of the coefficients. PROC PHREG also has a CLASS statement (in SAS 9.2 and later), but it has many more options than the one in PROC LIFEREG. PROC PHREG also has a TEST statement and a CONTRAST statement that make it easy to test any linear hypothesis about more than one coefficient. As we shall see, these two statements are somewhat redundant. Almost anything you can do with one can be done with the other. Furthermore, if you are already familiar with these statements from other SAS procedures (for example, REG, GLM, LOGISTIC, GENMOD), you won't find much new in this section.

As an example, let's treat the variable EDUC as a CLASS variable in a model for the recidivism data:

```
PROC PHREG DATA=recid;
   CLASS educ;
   MODEL week*arrest(0)=fin age prio educ/ TIES=EFRON;
   TEST educ3=educ5;
   CONTRAST 'ed3 vs. ed5' educ 0 1 0 -1 ;
RUN;
```

The EDUC variable has five different integer values: 2, 3, 4, 5, and 6. The CLASS statement creates a set of dummy variables to represent four out of the five categories. (In this context, they are sometimes referred to as *design variables*.) By default, the omitted (reference) category is the one with the highest value, in this case 6. You can override this default by specifying, for example,

```
CLASS educ(REF='2');
```

which makes level 2 the reference category.

The TEST statement tests the null hypotheses that the coefficient for level 3 is the same as the coefficient for level 5. Note that you refer to the coefficients by appending the variable value to the variable name. The CONTRAST statement accomplishes exactly the same task but with a very different syntax. The text enclosed in single quotes is a required label describing the test. It can be any text that you choose. The set of numbers following the variable name encodes the following instructions: Take the coefficient for the second dummy variable (EDUC3) and multiply it by 1. Take the coefficient for the fourth dummy variable (EDUC5) and multiply it by –1. Add the two numbers together and test whether the sum is equal to 0. Of course, this is equivalent to testing whether the two coefficients are equal to each other.

Output 5.26 displays the results. When a model contains a CLASS variable, PROC PHREG reports a Type 3 table. For variables that are not CLASS variables, this table is completely redundant with the usual regression table below. For the CLASS variable, however, it gives us the very useful test that all the coefficients associated with this variable are 0. What makes this test particularly attractive is that it does not depend on the choice of the reference category. In this case, with a chi-square of 4.53 and 4 degrees of freedom, the test is far from statistically significant.

In the regression table, we see coefficients and hazard ratios for levels 2 through 5 of EDUC. Each of these is a comparison with the omitted level 6. For example, the hazard ratio for level 3 is 2.738. This means that the hazard of arrest for someone at level 3 is almost three times the hazard for someone at level 6. However, the difference is not statistically significant, nor are any of the other differences with level 6. Below the regression table, we see the output from the TEST statement, which shows that the difference between level 3 and level 5 is not statistically significant. Below that is the same result from the CONTRAST statement.

You can include as many TEST or CONTRAST statements as you like. Also, a single CONTRAST or TEST statement can test more than one hypothesis at a time—just separate the hypotheses by commas. You'll get a single test statistic and *p*-value for a simultaneous test that all the hypotheses are true.

Output 5.26 *Results from Using the CLASS Statement for the Recidivism Data*

```
                                Type 3 Tests

                                    Wald
                    Effect     DF   Chi-Square    Pr > ChiSq

                    fin        1      3.6040        0.0576
                    age        1      8.1609        0.0043
                    prio       1      8.6708        0.0032
                    educ       4      4.5257        0.3395

                  Analysis of Maximum Likelihood Estimates

                     Parameter   Standard                      Hazard
    Parameter   DF    Estimate     Error  Chi-Square Pr > ChiSq  Ratio Label

    fin         1     -0.36314   0.19128    3.6040     0.0576    0.695
    age         1     -0.05986   0.02095    8.1609     0.0043    0.942
    prio        1      0.08355   0.02837    8.6708     0.0032    1.087
    educ   2    1      0.46192   1.12059    0.1699     0.6802    1.587 educ 2
    educ   3    1      1.00730   1.00996    0.9948     0.3186    2.738 educ 3
    educ   4    1      0.77558   1.01952    0.5787     0.4468    2.172 educ 4
    educ   5    1      0.28079   1.09597    0.0656     0.7978    1.324 educ 5

                  Linear Hypotheses Testing Results

                                 Wald
                    Label     Chi-Square     DF    Pr > ChiSq

                    Test 1      2.4159        1      0.1201

                       Contrast Test Results

                                     Wald
                    Contrast     DF   Chi-Square    Pr > ChiSq

                    ed3 vs. ed5   1     2.4159        0.1201
```

CUSTOMIZED HAZARD RATIOS

As we have seen, for each predictor variable PROC PHREG reports a hazard ratio (HR), which is just the exponentiated value of its coefficient. And if we further calculate 100(HR-1), we get the percentage change in the hazard of the event for a 1-unit increase in that particular variable. Two problems can arise, however. First, a 1-unit change may be either too small or too large to be usefully interpreted. If a person's income is measured in dollars, for example, a $1 increase will most likely produce a very tiny percentage change in the hazard. On the other hand, if a lab measurement varies only between 0 and 0.1, a 1-unit increase (from 0 to 1) will be 10 times the observed range of the variable. Although you can do hand calculations to get the hazard ratio for different units (or modify the units of the variables before estimating the model), the HAZARDRATIO statement (available in SAS 9.2 and later) makes things much easier.

Let's apply these options to the recidivism data. In Output 5.1, we saw a hazard ratio of .944 for AGE, which we interpreted to mean that each 1-year increase in age was associated with a 5.6 percent reduction in the hazard of arrest. Suppose, instead, that we want the hazard ratios for a 5-year increase and a 10-year increase in age. We can get that with the following code:

```
PROC PHREG DATA=my.recid;
   MODEL week*arrest(0)=fin age race wexp mar paro prio;
   HAZARDRATIO age / UNITS=5 10;
RUN;
```

In addition to the usual output, we now get Output 5.27, which tells us that a 5-year increase in age is associated with a 25 percent reduction in the hazard of arrest, while a 10-year increase in age is associated with a 44 percent reduction.

Output 5.27 *Results from Using the HAZARDRATIO Statement for the Recidivism Data*

Hazard Ratios for age			
Description	Point Estimate	95% Wald Confidence Limits	
age Unit=5	0.751	0.606	0.932
age Unit=10	0.564	0.367	0.868

The HAZARDRATIO statement also solves another problem. More recent versions of PROC PHREG allow for interactions to be directly specified in the MODEL statement. But if you include an interaction, PROC PHREG does not report hazard ratios for either the interaction itself or the main effects associated with the interaction. For example, suppose that we fit a model with an interaction between FIN and AGE, which produces the results in Output 5.28:

```
PROC PHREG DATA=my.recid;
   MODEL week*arrest(0)=fin age race wexp mar paro prio fin*age;
RUN;
```

Output 5.28 *Model with Interaction between FIN and AGE*

Parameter	DF	Parameter Estimate	Standard Error	Chi-Square	Pr > ChiSq	Hazard Ratio	Label
fin	1	1.62266	1.02829	2.4902	0.1146	.	
age	1	-0.02263	0.02595	0.7600	0.3833	.	
race	1	0.32112	0.30827	1.0851	0.2976	1.379	
wexp	1	-0.15996	0.21231	0.5677	0.4512	0.852	
mar	1	-0.45026	0.38264	1.3846	0.2393	0.637	
paro	1	-0.08408	0.19601	0.1840	0.6680	0.919	
prio	1	0.09380	0.02846	10.8618	0.0010	1.098	
fin*age	1	-0.08832	0.04504	3.8449	0.0499	.	fin * age

The interaction is just barely significant at the .05 level. Like all two-way interactions, this one can be interpreted in two different ways: the effect of AGE depends on the level of FIN and the effect of FIN depends on the level of AGE. The latter makes more sense because AGE is predetermined and FIN is an experimental treatment. So it is natural to ask how the effect of the treatment depends on the characteristics of the subjects. Output 5.28 is not helpful because it doesn't give us hazard ratios for either the interaction or the two main effects. To answer this question, we can use the HAZARDRATIO statement to give us hazard ratios for FIN at various selected ages. Here's how:

```
HAZARDRATIO fin / at (age=20 25 30 35 40) CL=PL;
```

The CL=PL option is not essential here. It requests that confidence intervals be computed using the profile likelihood method, which is somewhat more accurate than the conventional Wald method, especially in small samples.

Output 5.29 *Hazard Ratios for FIN by AGE*

```
┌─────────────────────────────────────────────────────────────┐
│                     Hazard Ratios for fin                    │
│                                                              │
│                                        95% Profile           │
│                              Point      Likelihood           │
│     Description            Estimate   Confidence Limits       │
│                                                              │
│     fin Unit=1 At age=20      0.866    0.559      1.333       │
│     fin Unit=1 At age=25      0.557    0.351      0.854       │
│     fin Unit=1 At age=30      0.358    0.157      0.743       │
│     fin Unit=1 At age=35      0.230    0.065      0.701       │
│     fin Unit=1 At age=40      0.148    0.027      0.676       │
└─────────────────────────────────────────────────────────────┘
```

Results are shown in Output 5.29. At age 20, the hazard ratio is not significantly different from 1 (as indicated by the 95% confidence interval, which includes 1). For later ages, all the odds ratios are significantly different from 1. The differences are striking. At age 25, receipt of financial aid reduces the hazard of arrest by 44 percent. At age 40, receipt of financial aid reduces the hazard by 85 percent.

BAYESIAN ESTIMATION AND TESTING

Beginning with SAS 9.2, PROC PHREG can do a Bayesian analysis of the Cox regression model. We discussed Bayesian analysis for PROC LIFEREG in the last section of Chapter 4. Because the theory, syntax, and output for Bayesian analysis are virtually identical for these two procedures, I will not go into all the details here.

As with PROC LIFEREG, requesting a Bayesian analysis in PROC PHREG can be as simple as including the BAYES statement after the MODEL statement. Here's how to do it for the recidivism data:

```
PROC PHREG DATA=my.recid;
  MODEL week*arrest(0)=fin age race wexp mar paro prio;
  BAYES;
RUN;
```

The default is to use non-informative priors for the coefficients and to produce 10,000 random draws from the posterior distribution using a Gibbs sampler algorithm. Output 5.30 displays some key summary statistics for the posterior distribution of the regression coefficients.

Output 5.30 *Results from Using the BAYES Statement for the Recidivism Data*

```
                              Bayesian Analysis

                           Posterior Summaries

                                   Standard              Percentiles
 Parameter        N       Mean    Deviation      25%         50%        75%

    fin        10000    -0.3823    0.1927      -0.5111    -0.3806    -0.2494
    age        10000    -0.0593    0.0220      -0.0738    -0.0586    -0.0443
    race       10000     0.3520    0.3133       0.1371     0.3434     0.5526
    wexp       10000    -0.1542    0.2113      -0.2988    -0.1534    -0.00844
    mar        10000    -0.4798    0.3959      -0.7338    -0.4593    -0.2022
    paro       10000    -0.0868    0.1947      -0.2180    -0.0887     0.0435
    prio       10000     0.0889    0.0286       0.0700     0.0897     0.1084


                           Posterior Intervals

   Parameter     Alpha      Equal-Tail Interval          HPD Interval

      fin        0.050     -0.7655    -0.00737      -0.7572    -0.00339
      age        0.050     -0.1047    -0.0186       -0.1028    -0.0175
      race       0.050     -0.2310     1.0003       -0.2634     0.9563
      wexp       0.050     -0.5708     0.2548       -0.5734     0.2499
      mar        0.050     -1.3054     0.2385       -1.2621     0.2575
      paro       0.050     -0.4669     0.3009       -0.4469     0.3194
      prio       0.050      0.0315     0.1435        0.0310     0.1425
```

As expected, the results from the Bayesian analysis do not differ much from the conventional analysis in Output 5.1. Either the means or medians (50th percentiles) of the posterior could be used as point estimates, and both are about the same as the conventional partial likelihood estimates. The standard deviations in Output 5.30 are about the same as the standard errors in Output 5.1. The big difference is in computing time, which increases by a factor of nearly 1,000 when the BAYES statement is used. There are a lot of additional tables in the default output that are not shown here, including fit statistics, autocorrelations, and Geweke diagnostics. And if you want to see graphs of the posterior distribution, you will need ODS statements before and after the code above, as shown in the last section of Chapter 4.

One surprising thing you can do with the BAYES statement is estimate a piecewise exponential model, which we discussed in Chapter 4 and estimated with PROC LIFEREG. For the piecewise exponential model, a special case of the Cox model, we divide the time scale into intervals with cut points $a_0, a_1, \ldots, a_J$, where $a_0 = 0$, and $a_J = \infty$. We assume that the hazard is constant within each interval but can vary across intervals. The hazard for individual i is assumed to have the form

$$\log h_i(t) = \alpha_j + \boldsymbol{\beta} \mathbf{x}_i \qquad \text{for } a_{j-1} \leq t < a_j$$

Thus, the intercept in the equation is allowed to vary in an unrestricted fashion from one interval to another.

You can do a Bayesian analysis of this model by putting the PIECEWISE option in the BAYES statement. On the other hand, if you only want maximum likelihood estimates, you can suppress the Gibbs sampler iterations with the statement

```
BAYES PIECEWISE NBI=0 NMC=1;
```

which gives the maximum likelihood estimates in Output 5.31. The default is to construct eight intervals with approximately equal numbers of events in each interval, as shown in the table. However, you can directly specify either the number of intervals or the cut points for the intervals.

Output 5.31 *Maximum Likelihood Estimates of Piecewise Exponential Model*

Constant Hazard Time Intervals				
Interval [Lower,	Upper)	N	Event	Log Hazard Parameter
0	11.5	17	17	Alpha1
11.5	18.5	16	16	Alpha2
18.5	24.5	15	15	Alpha3
24.5	30.5	12	12	Alpha4
30.5	36.5	14	14	Alpha5
36.5	42.5	13	13	Alpha6
42.5	47.5	13	13	Alpha7
47.5	Infty	332	14	Alpha8

(continued)

Output 5.31 *(continued)*

```
                   Maximum Likelihood Estimates

                                  Standard
     Parameter   DF    Estimate      Error     95% Confidence Limits

     Alpha1       1    -4.5804      0.6295     -5.8142      -3.3466
     Alpha2       1    -4.0752      0.6316     -5.3131      -2.8373
     Alpha3       1    -3.9299      0.6346     -5.1737      -2.6861
     Alpha4       1    -4.1037      0.6457     -5.3692      -2.8381
     Alpha5       1    -3.9063      0.6376     -5.1560      -2.6567
     Alpha6       1    -3.9233      0.6433     -5.1841      -2.6625
     Alpha7       1    -3.6959      0.6462     -4.9626      -2.4293
     Alpha8       1    -3.4703      0.6432     -4.7310      -2.2095
     fin          1    -0.3801      0.1914     -0.7552      -0.00507
     age          1    -0.0575      0.0220     -0.1006      -0.0144
     race         1     0.3131      0.3080     -0.2904       0.9167
     wexp         1    -0.1483      0.2123     -0.5645       0.2678
     mar          1    -0.4327      0.3820     -1.1813       0.3159
     paro         1    -0.0839      0.1957     -0.4675       0.2998
     prio         1     0.0917      0.0287      0.0355       0.1479
```

CONCLUSION

It's no accident that Cox regression has become the overwhelmingly favored method for doing regression analysis of survival data. It makes no assumptions about the shape of the distribution of survival times. It allows for time-dependent covariates. It is appropriate for both discrete-time and continuous-time data. It easily handles left truncation. It can stratify on categorical control variables. And, finally, it can be extended to nonproportional hazards. The principal *disadvantage* is that you lose the ability to test hypotheses about the shape of the hazard function. As we'll see in Chapter 8, however, the hazard function is often so confounded with unobserved heterogeneity that it's difficult to draw any substantive conclusion from the shape of the observed hazard function. So the loss may not be as great as it seems.

PROC PHREG goes further than most Cox regression programs in realizing the full potential of this method. In this chapter, I have particularly stressed those features of PROC PHREG that most distinguish it from other Cox regression programs: its extremely general capabilities for time-dependent covariates, its exact methods for tied data, and its ability to handle left truncation. If you already have some familiarity with PROC PHREG, you may have noticed that I said nothing about its several variable selection methods. I confess that I have never used them. While I am not totally opposed to such automated model-building methods, I think they should be reserved for those cases in which there are a large number of potential covariates, little theoretical guidance for choosing among them, and a goal that emphasizes prediction rather than hypothesis testing. If you find yourself in this situation, you may want to consider the best subset selection method that uses the SELECTION=SCORE and BEST= options. This method is computationally efficient because no parameters are estimated. Yet, it gives you lots of useful information.

We still haven't finished with PROC PHREG. Chapter 6, "Competing Risks," shows you how to use it for competing risks models in which there is more than one kind of event. Chapter 8, "Heterogeneity, Repeated Events, and Other Topics," considers PROC PHREG's capabilities for handling repeated events.

CHAPTER 6
Competing Risks

INTRODUCTION

In the previous chapters, all the events in each analysis were treated as if they were identical: all deaths were the same, all job terminations were the same, and all arrests were the same. In many cases, this is a perfectly acceptable way to proceed. But more often than not, it is essential—or at least desirable—to distinguish different kinds of events and treat them differently in the analysis. To evaluate the efficacy of heart transplants, you will certainly want to treat deaths due to heart failure differently from deaths due to accident or cancer. Job terminations that occur when an employee quits are likely to have quite different determinants than those that occur when the employee is fired. And financial aid to released convicts will more plausibly reduce arrests for theft or burglary than for rape or assault.

In this chapter, we consider the method of competing risks for handling these kinds of situations. What is most characteristic of competing risk situations is that the occurrence of one type of event removes the individual from risk of all the other event types. People who die of heart disease are no longer at risk of dying of cancer. Employees who quit can no longer be fired.

Because competing risk analysis requires no new SAS procedures, this chapter will be short. With only minor changes in the SAS statements, you can estimate competing risks models with the LIFETEST, LIFEREG, or PHREG procedures. All that's necessary is to change the code that specifies which observations are censored and which are not. After considering a bit of the theory of competing risks, most of the chapter will be devoted to examples. At the end, however, I also discuss some alternative to competing risks analysis that may be more appropriate for some situations.

Despite the practical importance of methods for handling multiple kinds of events, many textbooks give minimal coverage to this topic. For example Collett's (2003) otherwise excellent survey of survival analysis makes only a brief mention of competing risks. In this chapter, I have relied heavily on Kalbfleisch and Prentice (2002) and Cox and Oakes (1984).

TYPE-SPECIFIC HAZARDS

The classification of events into different types is often somewhat arbitrary and may vary according to the specific goals of the analysis. I'll have more to say about that later. For now, let's suppose that the events that we are interested in are deaths, and we have classified them into five types according to cause: heart disease, cancer, stroke, accident, and a residual category that we'll call *other*. Let's assign the numbers 1 through 5, respectively, to these death types. For each type of death, we are going to define a separate hazard function that we'll call a *type-specific* or *cause-specific hazard.*

As before, let T_i be a random variable denoting the time of death for person i. Now let J_i be a random variable denoting the type of death that occurred to person i. Thus, $J_5=2$ means that person 5 died of cancer. We now define $h_{ij}(t)$, the hazard for death type j at time t for person i, as follows:

$$h_{ij}(t) = \lim_{\Delta t \to 0} \frac{\Pr(t < T_i < t + \Delta t, J_i = j \mid T_i \geq t)}{\Delta t}, \quad j = 1,\ldots,5 \tag{6.1}$$

Comparing this with the definition of the hazard in equation (2.2), we see that the only difference is the appearance of $J_i = j$. Thus, the conditional probability in equation (6.1) is the probability that death occurs between t and $t+\Delta t$, *and* the death is of type j, given that the person had not already died by time t. The overall hazard of death is just the sum of all the type-specific hazards:

$$h_i(t) = \sum_j h_{ij}(t) \cdot \tag{6.2}$$

You can interpret type-specific hazards in much the same way as ordinary hazards. Their metric is the number of events per unit interval of time, except now the events are of a specific type. The reciprocal of the hazard is the expected length of time until an event of that type occurs, assuming that the hazard stays constant.

Based on the type-specific hazards, we can also define type-specific survival functions:

$$S_j(t) = \exp\left\{-\int_0^t h_j(u)du\right\}.$$ (6.3)

The interpretation of these functions is somewhat controversial, however. Lawless (2003) claims that they don't refer to any well-defined random variable and that they are useful only as a way of examining hypotheses about the hazard. On the other hand, we can define the hypothetical variable T_{ij} as the time at which the *j*th event type either occurred to the *i*th person *or would have occurred if other event types had not preceded it.* In other words, we suppose that a person who dies of cancer at time T_2 would have later died of heart disease at time T_1 if the cancer death had not occurred. For a given set of T_{ij}s, we only observe the one that is smallest. If we further assume that the T_{ij}s are independent across event types, we can say that

$$S_{ij}(t) = \Pr(T_{ij} > t).$$ (6.4)

That is, the type-specific survival function gives the probability that an event of type *j* occurs later than time *t*. You can see why people might want to forgo this interpretation, but I personally find it meaningful.

Now that we have the type-specific hazards, we can proceed to formulate models for their dependence on covariates. Any of the models that we have considered so far—both proportional hazards models and accelerated failure time models—are possible candidates. For example, we can specify general proportional hazards models for all five death types:

$$\log h_{ij}(t) = \alpha_j(t) + \boldsymbol{\beta}_j \mathbf{x}_i(t), \quad j = 1,\ldots,5.$$ (6.5)

where $\mathbf{x}_i(t)$ is a vector of covariates, some of which may vary with time. Note that the coefficient vector $\boldsymbol{\beta}$ is subscripted to indicate that the effects of the covariates may be different for different death types. In particular, some coefficients may be set to 0, thereby excluding the covariate for a specific death type. The $\alpha(t)$ function is also subscripted to allow the dependence of the hazard on time to vary across death types. If $\boldsymbol{\beta}_j$ and $\alpha_j(t)$ are the same for all *j*, then the model reduces to the proportional hazards model of Chapter 5 where no distinction is made among event types.

Although it is a bit unusual, there is nothing to prevent you from choosing, say, a log-normal model for heart disease, a gamma model for cancer, and a proportional hazards model for strokes. What makes this possible is that *the models may be estimated separately for each event type, with no loss of statistical precision.* This is, perhaps, the most important principle of competing risks analysis. More technically, the likelihood function for all event types taken together can be factored into a separate likelihood function for each event type. The likelihood function for each event type treats all other events as if the individual were censored at the time when the other event occurred.

There are a few computer programs that estimate models for competing risks simultaneously, but their only advantage is to reduce the number of statements needed to specify the models. This advantage must be balanced against the disadvantage of having to specify the same functional form and the same set of covariates for all event types.

The only time you need to estimate models for two or more events simultaneously is when there are parameter restrictions that cross event types. For example, Kalbfleisch and Prentice (2002) consider a model that imposes the restriction $\alpha_j(t) = \alpha_j + \alpha_0(t)$ on equation (6.3). This restriction says that the baseline hazard function for each event type is proportional to a common baseline hazard function. Later in the chapter, we'll see how to test this restriction (see **Estimates and Tests without Covariates**).

A further implication is that you don't need to estimate models for all event types unless you really want to. If you're only interested in the effects of covariates on deaths from heart disease, then you can simply estimate a single model for heart disease, treating all other death types as censoring. Besides reducing the amount of computation, this fact also makes it unnecessary to do an exhaustive classification of death types. You only need to distinguish the event type of interest from all other types of event.

Since we are treating events other than those of immediate interest as a form of censoring, it's natural to ask what assumptions must be made about the censoring mechanism. In Chapter 2, "Basic Concepts of Survival Analysis," we saw that censoring must be *noninformative* if the estimates are to be unbiased. We must make exactly the same assumption in the case of competing risks. That is, we must assume that, conditional on the covariates, those who are at particularly high (or low) risk of one event type are no more (or less) likely to experience other kinds of events. Thus, if we know that someone died of heart disease at age 50, that should give us no information (beyond what we know from the covariates) about his risk of dying of cancer at that age. *Noninformativeness* is implied by the somewhat stronger assumption that the hypothetical T_{ij}s (discussed above) are independent across j. Unfortunately, as we saw in Chapter 2 with censoring, it is impossible to test whether competing events are actually noninformative. I'll discuss this issue further in the conclusion to this chapter.

TIME IN POWER FOR LEADERS OF COUNTRIES: EXAMPLE

The main data set that I will use to illustrate competing risks analysis was constructed by Bienen and van de Walle (1991). They identified a fairly exhaustive set of primary leaders of all countries worldwide over the past 100 years or so. For each leader, they determined the number of years in power and the manner by which he or she lost power. Each period of leadership is called a *spell*. The mode of exit for a leader from a position of power was classified into three categories:

- death from natural causes
- constitutional means
- nonconstitutional means (including assassination).

Clearly, it is unreasonable to treat all three types as equivalent. While constitutional and nonconstitutional exits might have similar determinants, death from natural causes is clearly a distinct phenomenon.

The data set included the following variables:

YEARS	Number of years in power, integer valued. Leaders in power less than 1 year were coded 0.
LOST	0=still in power in 1987; 1=exit by constitutional means; 2=death by natural causes; and 3=nonconstitutional exit.
MANNER	How the leader reached power: 0=constitutional means; 1=nonconstitutional means.
START	Year of entry into power.
MILITARY	Background of the leader: 1=military; 0=civilian.
AGE	Age of the leader, in years, at the time of entry into power.
CONFLICT	Level of ethnic conflict: 1=medium or high; 0=low.
LOGINC	Natural logarithm of GNP per capita (dollar equivalent) in 1973.
GROWTH	Average annual rate of per capita GNP growth between 1965–1983.
POP	Population, in millions (year not indicated).
LAND	Land area, in thousands of square kilometers.
LITERACY	Literacy rate (year not indicated).
REGION	0=Middle East; 1=Africa; 2=Asia; 3=Latin America; and 4=North America, Europe, and Australia.

In the analysis that follows, I restrict the leadership spells to

- countries outside of Europe, North America, and Australia
- spells that began in 1960 or later
- only the first leadership spell for those leaders with multiple spells.

This leaves a total of 472 spells, of which 115 were still in progress at the time observation was terminated in 1987. Of the remaining spells, 27 ended when the leader died of natural causes, 165 were terminated by constitutional procedures, and 165 were terminated by nonconstitutional means. The restriction to starting years greater than 1960 was made so that the variables describing countries (income, growth, population, and literacy) would be reasonably concurrent with the exposure to risk.

ESTIMATES AND TESTS WITHOUT COVARIATES

The simplest question we might ask about the multiple event types is whether the type-specific hazard functions are the same for all event types (that is, $h_j(t) = h(t)$ for all j). For the LEADER data set, just looking at the frequencies of the three event types suggests that deaths due to natural causes ($n=27$) are much less likely to occur than the other two types, which have identical frequencies ($n=165$). In fact, we can easily obtain a formal test of the null hypothesis from these three frequencies. If the null hypothesis of equal hazards is correct, the expected frequencies of the three event types should be equal in any time interval. Thus, we have an expected frequency of $119=(472-115)/3$ for each of the three types. Calculating Pearson's chi-square test (by hand) yields a value of 88.9 with 2 d.f., which is far above the .05 critical value. So we can certainly conclude that the hazard for naturally occurring deaths is lower than for the other two exit modes.

Although the hazards are not equal, it's still possible that they might be proportional (in a different sense than that of Chapter 5); that is, if the hazard for death changes with time, the hazards for constitutional and nonconstitutional exits may also change by a proportionate amount. We can write this hypothesis as

$$h_j(t) = \omega_j h(t), \quad j = 1, 2, 3,$$ (6.6)

where the ω_js are constants of proportionality.

We can obtain a graphic examination of this hypothesis by using PROC LIFETEST to estimate log-log survivor functions for each of the three event types. If the hazards are proportional, the log-log survivor functions should be strictly parallel. We can get graphs of those functions by first creating a separate data set for each event type, with a variable EVENT equal to 1 if that event occurred and 0 otherwise. We then concatenate the three data sets into a single data set called COMBINE. Finally, we run PROC LIFETEST, stratifying on a variable TYPE that distinguishes the three data sets, and requesting the log-log survivor function with the PLOTS=LLS option. Here is the code:

```
DATA const;
  SET leaders;
  event=(lost=1);
  type=1;
DATA nat;
  SET leaders;
  event=(lost=2);
  type=2;
DATA noncon;
  SET leaders;
  event=(lost=3);
  type=3;
DATA combine;
  SET const nat noncon;
PROC LIFETEST DATA=COMBINE PLOTS=LLS;
  TIME years*event(0);
  STRATA type;
RUN;
```

Output 6.1 *Log-Log Survival Plot for Three Types of Exits, LEADERS Data Set*

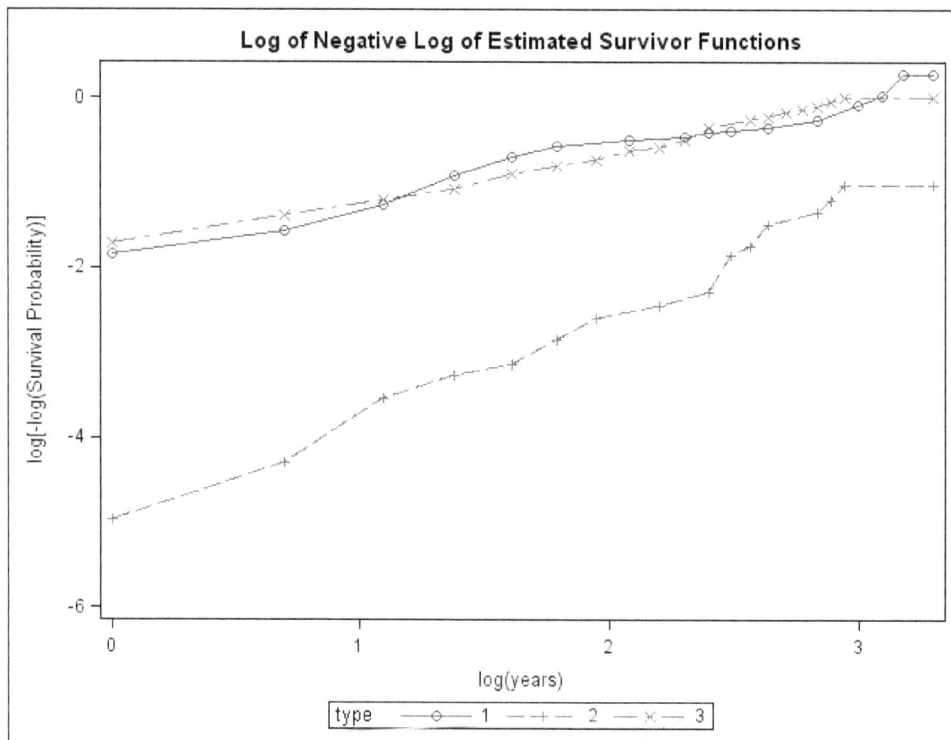

Output 6.1 shows the log-log survivor curves for the three event types. The curves for constitutional and nonconstitutional exits are virtually indistinguishable. Not surprisingly, the curve for natural deaths is much lower than the other two curves. There is also some tendency for the natural death curve to move closer to the other two in later years, which is evidence against the proportionality hypothesis.

We can also examine smoothed hazard plots using the kernel smoothing option for PROC LIFETEST that was described in Chapter 3:

```
ODS GRAPHICS ON;
PROC LIFETEST DATA=combine PLOTS=H(BW=10);
  TIME years*event(0);
  STRATA type;
RUN;
ODS GRAPHICS OFF;
```

In Output 6.2, we see that the smoothed hazard functions for constitutional and nonconstitutional exits are very similar, except after the 15th year. The sharp increase in the hazard for constitutional exits after this year could arguably be disregarded because the standard errors also become large at later points in time. On the other hand, it could represent the fact that many governments have constitutional limits on the number of years a leader can spend in power. As might be expected, the hazard for natural death is much lower but gradually increases with time (except for the very end where, again, the estimates are not very reliable).

Output 6.2 *Smoothed Hazard Functions for Three Types of Exits*

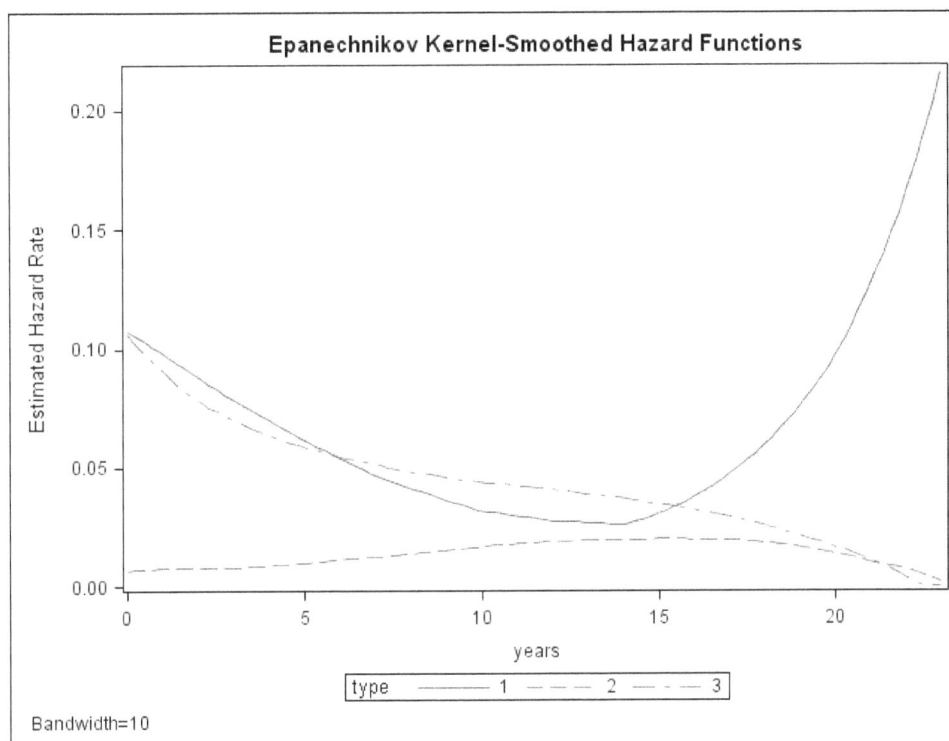

Cox and Oakes (1984) proposed a parametric test of the proportional hazards hypothesis in equation (6.6). Consider the model

$$\log h_j(t) = \alpha_0(t) + \alpha_j + \beta_j t, \quad j = 1, 2, \ldots \tag{6.7}$$

If $\beta_j = \beta$ for all j, then the proportional hazards hypothesis is satisfied. Otherwise, this model says that the log-hazards for any two event types diverge linearly with time. Cox and Oakes showed that if there are two event types, equation (6.7) implies a binary logistic regression model for the type of event, with the time of the event as an independent variable. Under the proportional hazards hypothesis, the coefficient for time will be 0.

For more than two event types, equation (6.7) implies a *multinomial* logit model for event type. Here's how to estimate that model with PROC LOGISTIC:

```
PROC LOGISTIC DATA=leaders;
   WHERE lost NE 0;
   MODEL lost=years / LINK=GLOGIT;
RUN;
```

Notice that leadership spells that are still in progress (LOST=0) when the observation is terminated are excluded from the analysis. The LINK=GLOGIT option specifies an unordered multinomial logit model (rather than the default cumulative logit model). Output 6.3 shows selected portions of the output.

Output 6.3 *Test of Proportionality with PROC LOGISTIC*

```
                    Type 3 Analysis of Effects

                                  Wald
              Effect      DF   Chi-Square    Pr > ChiSq

              years        2     17.8915       0.0001


            Analysis of Maximum Likelihood Estimates

                                    Standard       Wald
Parameter    lost   DF   Estimate     Error    Chi-Square   Pr > ChiSq

Intercept     1      1     0.0134     0.1432      0.0088       0.9253
Intercept     2      1    -2.5393     0.3141     65.3689      <.0001
years         1      1    -0.00391    0.0267      0.0215       0.8834
years         2      1     0.1394     0.0359     15.0990       0.0001
```

Looking first at the TYPE 3 table, we see that the effect of YEARS is highly significant, indicating a rejection of the proportionality hypothesis. The coefficients in the lower half of the output tell us which hazard functions are proportional and which are not. The first coefficient for YEARS is the β coefficient for the contrast between type 1 (constitutional exit) and type 3 (nonconstitutional exit). The chi-square statistic is minuscule, indicating that proportionality cannot be rejected for these two hazard types. Moreover, the fact that the coefficient is also near 0 suggests that the hazard functions for these two event types are nearly identical. On the other hand, the second YEARS coefficient is highly significant. This coefficient is a contrast between the type 2 hazard (death from natural causes) and the type 3 hazard (nonconstitutional exit). Its value tells us that the hazard for natural death increases much more rapidly with time than the hazard for nonconstitutional exit; specifically, their ratio increases by about 15 percent each year (calculated as $100(\exp(.1394)-1)$)).

COVARIATE EFFECTS VIA COX MODELS

We have just seen evidence that the hazard for natural death is lower than that for the other two exit types but that it increases more rapidly with time. On the other hand, the hazard functions for constitutional and nonconstitutional exits are nearly indistinguishable. We now look at whether the effects of covariates are the same or different across event types by fitting a Cox model to each type. The SAS program for doing this is as follows:

```
PROC PHREG DATA=leaders;
   MODEL years*lost(0)=manner start military age conflict
        loginc growth pop land literacy;
   STRATA region;
PROC PHREG DATA=leaders;
   MODEL years*lost(0,1,2)=manner start military age conflict
        loginc growth pop land literacy;
   STRATA region;
PROC PHREG DATA=leaders;
   MODEL years*lost(0,1,3)=manner start military age conflict
        loginc growth pop land literacy;
   STRATA region;
PROC PHREG DATA=leaders;
   MODEL years*lost(0,2,3)=manner start military age conflict
        loginc growth pop land literacy;
   STRATA REGION;
RUN;
```

The first model treats all event types the same. I've included that model for comparison, as well as for some hypothesis tests that we'll look at shortly. The second model focuses on type 3 by treating types 1 and 2 as censoring. That's followed by a model for type 2, treating types 1 and 3 as censoring, and a model for type 1, treating types 2 and 3 as censoring. Notice that I have stratified by region. Alternatively, I could have created three dummy variables to represent the four regions and included them in the model. However, stratification is less restrictive because it allows for interactions between region and time, as we saw in Chapter 5.

In Output 6.4 we see what happens when no distinction is made among the different kinds of exits. Four variables are statistically significant by conventional criteria. The strongest effect is age at entry into power, with each additional year of age associated with a 2 percent increase in the risk of leaving power. Recall from Chapter 5, moreover, that if age at origin is a covariate in a Cox model, its coefficient may actually be interpreted as the effect of age as a time-dependent covariate. We also see that

- leaders who attained power by nonconstitutional means (MANNER=1) have a 44 percent greater risk of leaving power
- leaders in countries with higher GNP per capita (LOGINC) have a lower risk of exit
- the risk of exit declined by about 1.6 percent per year (START) since 1960.

The coefficient for the logarithm of GNP (−.1710) needs a little explanation. If both the hazard and a covariate in a regression model are logged, we can interpret the coefficient (without transformation) as the percent change in the hazard for a *1 percent* increase in the covariate (in its original metric). Thus, we conclude that a 1 percent increase in per capita GNP yields a .17 percent decrease in the hazard of exit.

Output 6.4 *PROC PHREG Analysis for Exits by Any Means*

```
                    Testing Global Null Hypothesis: BETA=0

          Test                 Chi-Square      DF      Pr > ChiSq

          Likelihood Ratio       29.9069       10        0.0009
          Score                  29.5835       10        0.0010
          Wald                   29.3945       10        0.0011


                 Analysis of Maximum Likelihood Estimates

                      Parameter   Standard                          Hazard
    Parameter   DF     Estimate      Error   Chi-Square  Pr > ChiSq  Ratio

    manner       1      0.36403    0.15412      5.5787     0.0182    1.439
    start        1     -0.01591    0.00815      3.8129     0.0509    0.984
    military     1     -0.22510    0.16251      1.9186     0.1660    0.798
    age          1      0.02016    0.00555     13.1731     0.0003    1.020
    conflict     1      0.12052    0.13175      0.8368     0.3603    1.128
```

(*continued*)

Output 6.4 *(continued)*

loginc	1	-0.17103	0.08244	4.3044	0.0380	0.843
growth	1	-0.00140	0.02131	0.0043	0.9478	0.999
pop	1	-0.0000210	0.0006375	0.0011	0.9737	1.000
land	1	0.0000107	0.0000478	0.0502	0.8228	1.000
literacy	1	0.0008773	0.00321	0.0747	0.7847	1.001

The picture changes somewhat when we focus on the hazard for nonconstitutional exits in Output 6.5. Leaders who acquired power by nonconstitutional means are 2.4 times as likely as other leaders to exit by nonconstitutional means. Income also has a stronger effect: a 1 percent increase in per capita GNP now yields a .42 percent decrease in the risk of nonconstitutional exit. Age is no longer important, but we still see a reduction in the hazard over time since 1960. In addition, we find that leaders in countries with ethnic conflict have a 62 percent greater chance of nonconstitutional exit.

Constitutional exits show a different pattern (Output 6.6). Manner of acquiring power is not important, nor is income or conflict. Age shows up again, however, and we also see an effect of literacy. Each 1-percentage-point increase in the literacy rate is associated with a 1.4 percent increase in the risk of a constitutional exit.

Output 6.5 *PROC PHREG Analysis for Nonconstitutional Exits*

Parameter	DF	Parameter Estimate	Standard Error	Chi-Square	Pr > ChiSq	Hazard Ratio
manner	1	0.88049	0.21958	16.0796	<.0001	2.412
start	1	-0.03189	0.01220	6.8278	0.0090	0.969
military	1	-0.38832	0.22651	2.9392	0.0865	0.678
age	1	0.00827	0.00841	0.9670	0.3254	1.008
conflict	1	0.48053	0.20398	5.5498	0.0185	1.617
loginc	1	-0.42010	0.14129	8.8412	0.0029	0.657
growth	1	-0.04585	0.03130	2.1466	0.1429	0.955
pop	1	-0.00101	0.00157	0.4170	0.5184	0.999
land	1	0.0000185	0.0000809	0.0525	0.8188	1.000
literacy	1	-0.00572	0.00454	1.5901	0.2073	0.994

Output 6.6 *PROC PHREG Analysis for Constitutional Exits*

Parameter	DF	Parameter Estimate	Standard Error	Chi-Square	Pr > ChiSq	Hazard Ratio
manner	1	-0.27001	0.25623	1.1105	0.2920	0.763
start	1	0.00276	0.01200	0.0530	0.8178	1.003
military	1	-0.02793	0.25913	0.0116	0.9142	0.972
age	1	0.02267	0.00865	6.8729	0.0088	1.023
conflict	1	-0.03383	0.20309	0.0278	0.8677	0.967
loginc	1	-0.12894	0.11904	1.1733	0.2787	0.879
growth	1	0.03254	0.03452	0.8887	0.3458	1.033
pop	1	0.0004148	0.0008444	0.2413	0.6233	1.000
land	1	-0.0000233	0.0000699	0.1108	0.7393	1.000
literacy	1	0.01376	0.00559	6.0532	0.0139	1.014

For deaths from natural causes (Output 6.7), it is not terribly surprising that age is the only significant variable. Each 1-year increase in age is associated with a 7.8 percent increase in the hazard of death.

Output 6.7 *Deaths from Natural Causes*

Parameter	DF	Parameter Estimate	Standard Error	Chi-Square	Pr > ChiSq	Hazard Ratio
manner	1	0.30550	0.70173	0.1895	0.6633	1.357
start	1	-0.05731	0.03526	2.6410	0.1041	0.944
military	1	-0.30438	0.78263	0.1513	0.6973	0.738
age	1	0.07834	0.02035	14.8145	0.0001	1.081
conflict	1	-0.55599	0.50965	1.1901	0.2753	0.574
loginc	1	0.19416	0.28346	0.4692	0.4934	1.214
growth	1	0.09201	0.08551	1.1579	0.2819	1.096
pop	1	0.0009613	0.00220	0.1910	0.6621	1.001
land	1	0.0000348	0.0001800	0.0373	0.8468	1.000
literacy	1	-0.01268	0.01364	0.8644	0.3525	0.987

We see, then, that the coefficients can differ greatly across different event types. But perhaps these differences are merely the result of random variation. What we need is a test of the null hypothesis that $\beta_j = \beta$ for all j, where β_j is the vector of coefficients for event type j. A test statistic is readily constructed from output given by PROC PHREG. For each model,

PROC PHREG reports $-2 \times$ log-likelihood (for the model with covariates). For the four models we just estimated, the values are

All types combined	2620.54
Nonconstitutional	1216.76
Constitutional	1158.92
Natural death	156.24

To construct the test, we sum the values for the three specific death types, yielding a total of 2531.92. We then subtract that from the value for all types combined, for a difference of 88.62. This is a likelihood ratio chi-square statistic for the null hypothesis. How many degrees of freedom? Well, when we estimated separate models for the three event types, we got a total of 30 coefficients. When we collapsed them all together, we only estimated 10. The difference of 20 is the degrees of freedom. Because the chi-square statistic is significant at well beyond the .01 level, we may reject the hypothesis that the coefficients are all equal across event types.

That result is not terribly surprising because deaths from natural causes are unlikely to have the same determinants as constitutional and nonconstitutional exits. A more interesting question is whether the covariates for the latter two event types have identical coefficients. We can obtain a test statistic for that null hypothesis by estimating a model that combines constitutional and nonconstitutional exits and that treats natural deaths as censoring. For that model, $-2 \times$ log-likelihood=2443.71. If we sum the values for the two separate models (given above), we get 2375.68. The difference of 68.03 has 10 degrees of freedom (the difference in the number of estimated parameters) for a *p*-value less than .0001. So again, we may conclude that different models are required for constitutional and nonconstitutional exits.

These tests (for all three events or for just two of them) are only valid when using the Breslow method for handling ties or when there are no tied event times. A more complicated procedure is necessary if there are ties, and one of the other three tie-handling methods is used. Suppose, for example, that we re-estimate all the models using the TIES=EFRON option in the MODEL statement. We get the following values for the -2 log L statistics:

All types combined	2561.08
Nonconstitutional	1202.04
Constitutional	1139.41
Natural death	155.38

As before, we can add up the -2 log L statistics for the three separate event types, yielding a total of 2496.83. But we can't validly compare this number with –2 log L for all the types combined. That's because when we estimate three models for the different outcomes, we are allowing the baseline hazard function to be different for each outcome. When we estimate the model for all the types combined, however, we are only allowing for a single baseline hazard function. The usual likelihood ratio chi-square has no way to adjust for this difference.

The solution is to use the COMBINE data set that was created in the **Estimation and Testing without Covariates** section above. This data set was constructed by first creating a separate data set for each event type, with a variable EVENT equal to 1 if that event occurred and 0 otherwise. We then concatenated the three data sets into a single data set (COMBINE), which included the variable TYPE with values of 1, 2, or 3 to distinguish the three kinds of deaths. We now estimate a Cox model for this data set while stratifying on the TYPE variable. We are already stratifying by REGION so we can simply add TYPE to the STRATA statement:

```
PROC PHREG DATA=combine;
   MODEL years*event(0)=manner start military age conflict
         loginc growth pop land literacy / TIES=EFRON;
   STRATA region type;
RUN;
```

This produces a –2 log L of 2590.68. From this we subtract the summed –2 log Ls for the three separate models (2496.83) to produce a likelihood ratio chi-square of 93.85. With 20 degrees of freedom, the chi-square statistic is highly significant, leading again to the conclusion that the regression coefficients differ across the three event types.

If we had concluded, instead, that corresponding coefficients were equal across all event types, a natural next step would be to test whether they were all equal to 0. No special computations are needed for that test, however. Just estimate the model without distinguishing among the event types, and examine the global statistics for the null hypothesis that all coefficients are equal to 0. In Output 6.4, all three chi-square statistics (Wald, score, and likelihood ratio) have values of about 30 with 10 d.f., giving strong evidence against that null hypothesis. Note that this is a conditional test, the condition being that the coefficients are equal. To test the hypothesis that corresponding coefficients are equal *and* that they are equal to 0, simply add this conditional chi-square statistic to the chi-square statistic for testing equality and then add the degrees of freedom as well. Thus, for this example, the likelihood ratio statistic is 29.90+88.62=118.52 with 40 d.f.

We can also construct test statistics for hypotheses about coefficients for specific covariates, using only the coefficients and their standard errors

(Lagakos, 1978). The coefficient for CONFLICT for nonconstitutional exits was .4805, with a standard error of .2040. By contrast, the coefficient for CONFLICT for constitutional exits was −.0338, with a standard error of .2031. The first coefficient is significantly different from 0; the second is not. But is there a significant difference between them? A 1-degree-of-freedom Wald chi-square statistic for testing the null hypothesis that $\beta_1 = \beta_2$ is easily calculated by the following formula:

$$\frac{\left(b_1 - b_2\right)^2}{\left[s.e.(b_1)\right]^2 + \left[s.e.(b_2)\right]^2} \tag{6.8}$$

where b_1 is the estimate of β_1 and $s.e.(.)$ means estimated standard error. For our particular question, we have

$$\frac{\left(-.03383 - .48053\right)^2}{(.2031)^2 + (.2040)^2} = 3.19 \,.$$

Because that does not exceed the .05 critical value, we have insufficient statistical justification for concluding that the two coefficients are different.

This particular statistic, in one form or another, is widely used to test for differences in parameter estimates across two independent groups. Here, however, we do not have independent groups because the same set of 472 leaders is used to estimate models for both constitutional and nonconstitutional exits. This suggests that we may need a covariance term in the denominator of the statistic. In fact, we do not. What justifies this statistic is the fact that the likelihood function factors into a distinct likelihood for each event type. It follows that the parameter estimates for each event type are asymptotically independent of the parameter estimates for all other event types. This only holds for mutually exclusive event types, however. We cannot use this statistic to test for the difference between a coefficient for constitutional exits and the corresponding coefficient for all types of exits.

Because we found no reason to reject the hypothesis that the two coefficients for CONFLICT are equal, we may want to go further and test whether they are 0. This is easily accomplished by taking the reported chi-square statistics for CONFLICT, for both constitutional and nonconstitutional exits, and summing them: .87+5.55=6.42 with 2 d.f. This is just barely significant at the .05 level.

The tests for a single covariate can be generalized to more than two event types. For a given covariate, let b_j be its coefficient for event type j; let s_j^2 be the squared, estimated standard error of b_j; and let $X_j^2 = b_j^2 / s_j^2$ be the reported Wald chi-square statistic for testing that $\beta_j = 0$. To test the hypothesis that all the coefficients for the chosen covariate are 0, we sum the Wald chi-square statistics as follows:

$$Q = \sum_j X_j^2 \tag{6.9}$$

which has degrees of freedom equal to the number of event types. To test the hypothesis that all coefficients are equal to each other, we calculate

$$Q - \frac{\left(\sum_j \dfrac{b_j}{s_j^2}\right)^2}{\sum_j \dfrac{1}{s_j^2}} \tag{6.10}$$

which has degrees of freedom equal to one less than the number of event types.

ACCELERATED FAILURE TIME MODELS

Competing risks analysis with accelerated failure time models is basically the same as with Cox models. As before, the key point is to treat all events as censoring except the one that you're focusing on. However, there are some complications that arise in constructing tests of equality of coefficients across event types. There are also some special characteristics of the LEADER data set that require slightly different treatment with accelerated failure time models.

One of those characteristics is the presence of 0s in the YEARS variable. Recall that any leaders who served less than 1 year were assigned a time value of 0. Of the 472 leaders, 106 fell in this category. This poses no problem for PROC PHREG, which is only concerned with the rank order of the time variable. On the other hand, PROC LIFEREG excludes any observations with times of 0 or less because it must take the logarithm of the event time as the first step. But we certainly don't want to exclude 22 percent of the cases. One approach is to assign some arbitrarily chosen number between 0 and 1. A more elegant solution is to treat such cases as if they were left censored at time 1. We discussed how to do this in some detail in the section **Left Censoring and Interval Censoring** in Chapter 4, "Estimating Parametric Regression Models with PROC LIFEREG." For the LEADER data set, we need a short DATA step to create LOWER and UPPER variables that are appropriately coded. Here's how to prepare the data set for modeling the constitutional exits.

```
DATA leaders2;
   SET leaders;
   lower=years;
   upper=years;
   IF years=0 THEN DO;
      lower=.;
      upper=1;
   END;
   IF lost IN (0,1,2) THEN upper=.;
RUN;
```

For uncensored observations, LOWER and UPPER have the same value. For observations with YEARS=0, we set LOWER=. and UPPER=1. For right-censored observations (including those with events other than the one of interest), UPPER=. and LOWER=YEARS. If both UPPER and LOWER are missing (which happens for individuals with YEARS=0 who have events other than the one of interest), the observation is excluded. To fit a model for the other two types of exit, we simply change the last IF statement so that the numbers in parentheses include all outcomes to be treated as right censored.

To fit an exponential model to this data set, I used the following PROC LIFEREG program:

```
PROC LIFEREG DATA=leaders2;
   CLASS region;
   MODEL (lower,upper)= manner start military age conflict
         loginc literacy region / D=EXPONENTIAL;
RUN;
```

Notice that REGION is declared to be a CLASS variable. In the PROC PHREG analysis, I simply stratified on this variable. But stratification is not an option in PROC LIFEREG. I also excluded from the models all variables that were not statistically significant in any of the PROC PHREG models (namely, POP, GROWTH, and LAND).

For each of the three event types, I estimated all five of the accelerated failure time models discussed in Chapter 4. The models and their log-likelihoods are shown below:

	Nonconstitutional	Constitutional	Natural Death
Exponential	-383.39	-337.30	-87.17
Weibull	-372.51	-336.46	-82.48
Log-normal	-377.04	-338.09	-83.60
Gamma	-372.47	-336.14	(-81.36)
Log-logistic	-374.95	-335.88	-82.78

(In fitting the gamma model to natural deaths, PROC LIFEREG reported a possible convergence failure after 22 iterations. I've given the log-likelihood at the last iteration.)

For the nonconstitutional exits, we can reject the hypothesis that the hazard is constant over time. That's because the chi-square statistic for comparing the exponential (constant hazard) model with the generalized gamma model is 21.84, which was calculated by taking the difference in the log-likelihoods for the two models and multiplying by 2. With 2 d.f., this has a *p*-value of .00002. The log-normal model must also be rejected (*p*=.003). The Weibull model, on the other hand, is only trivially different from the gamma model.

Output 6.8 shows the parameter estimates for the Weibull model. Consider first the scale estimate of 1.41, which can tell us whether the hazard is increasing or decreasing. Using the transformation 1/1.41−1= −.29, we get the coefficient of log *t* in the equivalent proportional hazards model. This result indicates that the hazard of a nonconstitutional exit *decreases* with time since entry into power. The *p*-values for the covariates are quite similar to those in Output 6.5 for the Cox model, but naturally the coefficients are all reversed in sign. To directly compare the magnitudes of the coefficients with those of the Cox model, each must be divided by the scale estimate of 1.41. After this adjustment, the values are remarkably similar. We also see major differences among the regions, something that was hidden in the Cox analysis. Specifically, expected time until a nonconstitutional exit is more than seven times greater (exp(2.036)) in Asia (REGION=2) than it is in Latin America (REGION=3), and it is nearly four times greater (exp(1.36)) in Africa (REGION=1).

Output 6.8 *Weibull Model for Nonconstitutional Exits*

Parameter	DF		Estimate	Standard Error	95% Confidence Limits		Chi-Square	Pr > ChiSq
Intercept		1	-40.6136	17.1902	-74.3057	-6.9215	5.58	0.0181
manner		1	-1.3806	0.3143	-1.9965	-0.7646	19.30	<.0001
start		1	0.0410	0.0178	0.0062	0.0758	5.33	0.0210
military		1	0.6510	0.3135	0.0366	1.2655	4.31	0.0378
age		1	-0.0124	0.0114	-0.0348	0.0100	1.18	0.2775
conflict		1	-0.7199	0.2843	-1.2771	-0.1626	6.41	0.0113
loginc		1	0.6750	0.2073	0.2688	1.0812	10.61	0.0011
literacy		1	0.0072	0.0063	-0.0052	0.0196	1.28	0.2572
region	0	1	0.9037	0.4457	0.0301	1.7773	4.11	0.0426
region	1	1	1.3636	0.3926	0.5941	2.1332	12.06	0.0005

(continued)

Output 6.8 *(continued)*

```
region          2  1    2.0362   0.4692   1.1166   2.9558   18.83    <.0001
region          3  0    0.0000     .        .        .        .        .
Scale              1    1.4064   0.1121   1.2030   1.6441
Weibull Shape      1    0.7111   0.0567   0.6082   0.8313
```

For constitutional exits, the picture is rather different. All the models have similar log-likelihoods, and even the exponential model is not significantly worse than the Weibull or gamma models. Invoking parsimony, we might as well stick with the exponential model, which is displayed in Output 6.9. The Lagrange multiplier test at the bottom of the output shows, again, that the exponential model cannot be rejected. As with the nonconstitutional exits, the results are quite similar to those of the Cox model in Output 6.6. Apart from the region differences, the only significant variables are AGE and LITERACY. Except for the reversal of sign, the coefficients are directly comparable to those of the Cox model. Again, we see that Latin America has the shortest expected time until a constitutional exit, but the longest expected time is now in Africa rather than Asia.

Output 6.9 *Exponential Model for Constitutional Exits*

Parameter		DF	Estimate	Standard Error	95% Confidence Limits		Chi-Square	Pr > ChiSq
Intercept		1	12.4138	11.7600	-10.6353	35.4630	1.11	0.2912
manner		1	0.3112	0.2624	-0.2030	0.8255	1.41	0.2356
start		1	-0.0092	0.0120	-0.0328	0.0144	0.58	0.4453
military		1	0.0284	0.2526	-0.4667	0.5236	0.01	0.9104
age		1	-0.0317	0.0083	-0.0480	-0.0154	14.52	0.0001
conflict		1	0.1573	0.1983	-0.2314	0.5460	0.63	0.4277
loginc		1	0.1366	0.1135	-0.0858	0.3590	1.45	0.2287
literacy		1	-0.0113	0.0056	-0.0223	-0.0003	4.09	0.0432
region	0	1	0.5413	0.3326	-0.1106	1.1932	2.65	0.1036
region	1	1	1.7032	0.3702	0.9776	2.4288	21.17	<.0001
region	2	1	0.5336	0.2180	0.1062	0.9609	5.99	0.0144
region	3	0	0.0000	.	.	.	.	.
Scale		0	1.0000	0.0000	1.0000	1.0000		
Weibull Shape		0	1.0000	0.0000	1.0000	1.0000		

Lagrange Multiplier Statistics

Parameter	Chi-Square	Pr > ChiSq
Scale	1.4704	0.2253

For deaths due to natural causes, the only model that is clearly rejectable is the exponential model. Of the remaining models, the Weibull model again appears to be the best choice because its log-likelihood is only trivially different from that of the gamma model. The Weibull estimates are reported in Output 6.10. As in the Cox model, there is a strong effect of age at entry, with older leaders having shorter times until death.

Output 6.10 *Weibull Estimates for Natural Deaths*

Parameter		DF	Estimate	Standard Error	95% Confidence Limits		Chi-Square	Pr > ChiSq
Intercept		1	-23.1476	20.0574	-62.4595	16.1642	1.33	0.2485
manner		1	-0.2216	0.3909	-0.9878	0.5445	0.32	0.5707
start		1	0.0302	0.0209	-0.0107	0.0712	2.09	0.1480
military		1	0.2210	0.4158	-0.5939	1.0360	0.28	0.5950
age		1	-0.0451	0.0108	-0.0662	-0.0240	17.58	<.0001
conflict		1	0.0736	0.2804	-0.4761	0.6232	0.07	0.7930
loginc		1	-0.1514	0.1563	-0.4577	0.1549	0.94	0.3328
literacy		1	0.0030	0.0073	-0.0112	0.0172	0.17	0.6769
region	0	1	0.3597	0.4472	-0.5168	1.2361	0.65	0.4212
region	1	1	0.7672	0.4491	-0.1131	1.6475	2.92	0.0876
region	2	1	0.6378	0.3745	-0.0963	1.3718	2.90	0.0886
region	3	0	0.0000	.	.	.	.	.
Scale		1	0.5994	0.0885	0.4487	0.8007		
Weibull Shape		1	1.6684	0.2465	1.2489	2.2287		

The next logical step is to construct tests of hypotheses about equality of coefficients across different event types. For tests about individual covariates, the chi-square statistics in equations (6.8) – (6.10) will do just fine. Unfortunately, we cannot use the method employed for Cox models to construct a global test of whether all the coefficients for one event type are equal to the corresponding coefficients for another event type. For the accelerated failure time models, fitting a model to all events without distinction involves a different likelihood than constraining parameters to be equal across the separate likelihoods. As a result, if you try to calculate the statistics described in the previous section, you're likely to get negative chi-square statistics.

In general, there's no easy way around this problem. For Weibull models (actually for any parametric proportional hazards model), the likelihood test can be corrected by a function of the number of events of each type at each point in time (Narendranathan and Stewart, 1991). Here, I consider an alternative test for Weibull models that I think is more easily performed. It also yields a test for each covariate as a by-product. It is, in fact, a simple generalization of the test for proportional hazards discussed earlier in this chapter (see **Time in Power for Leaders of Countries: Example**).

Suppose we believe that constitutional and nonconstitutional exits are both governed by Weibull models but with different parameters. We can write the two models as

$$\log h_j(t) = \alpha_j \log t + \beta_{0j} + \beta_{1j} x_1 + \ldots + \beta_{kj} x_k \tag{6.11}$$

with $j=1$ for constitutional exits and $j=2$ for nonconstitutional exits. Now consider the following question: Given that an exit (other than a natural death) occurs at time t, what determines whether it is a constitutional or a nonconstitutional exit? Equation (6.11) implies that this question is answered by a logistic regression model:

$$\log \frac{\Pr(J=1 \mid T=t)}{\Pr(J=2 \mid T=t)} = (\alpha_1 - \alpha_2)\log t + (\beta_{01} - \beta_{02}) + (\beta_{11} - \beta_{12})x_1 + \ldots + (\beta_{k1} - \beta_{k2})x_k$$

This model can be easily estimated with PROC LOGISTIC:

```
DATA leaders3;
   SET leaders;
   lyears=LOG(years+.5);
PROC LOGISTIC DATA=leaders3;
   WHERE lost=1 OR lost=3;
   CLASS region / PARAM=GLM;
   MODEL lost=lyears manner age start military conflict loginc
         literacy region;
RUN;
```

The DATA step is needed so that the log of time can be included in the model (0.5 was added to avoid problems with times of 0). The WHERE statement eliminates the censored cases and those who died of natural causes.

Output 6.11 shows the results. We can interpret each of the coefficients in this table as an estimate of the difference between corresponding coefficients in the two Weibull models. We see that there are highly significant differences in the coefficients for MANNER and LITERACY but only marginally significant differences between the coefficients for AGE, START, and LYEARS. The test for LYEARS is equivalent to a test of whether the scale parameters are the same in the accelerated failure time version of the model.

The null hypothesis that *all* the corresponding coefficients are equal in the two Weibull models is equivalent to the hypothesis that all the coefficients in the implied logit model are 0. That hypothesis, of course, is tested by the three chi-square statistics in the panel labeled "Testing Global Null Hypothesis: BETA=0". Although the three statistics vary quite a bit in magnitude, they all indicate that the hypothesis should be rejected.

Output 6.11 *Logit Model Comparing Constitutional and Nonconstitutional Exits*

```
              Testing Global Null Hypothesis: BETA=0

        Test                 Chi-Square       DF      Pr > ChiSq

        Likelihood Ratio      129.1322        11        <.0001
        Score                 108.7634        11        <.0001
        Wald                   76.2327        11        <.0001


                Analysis of Maximum Likelihood Estimates

                                      Standard        Wald
        Parameter      DF    Estimate    Error    Chi-Square    Pr > ChiSq

        Intercept       1    -42.9260   20.7274       4.2890      0.0384
        lyears          1      0.2749    0.1424       3.7239      0.0536
        manner          1     -1.3252    0.3801      12.1579      0.0005
        age             1      0.0277    0.0144       3.7136      0.0540
        start           1      0.0407    0.0214       3.6004      0.0578
        military        1      0.0698    0.3949       0.0313      0.8596
        conflict        1     -0.2433    0.3425       0.5044      0.4776
        loginc          1      0.0581    0.2502       0.0539      0.8163
        literacy        1      0.0334   0.00895      13.9201      0.0002
        region    0     1      0.1111    0.5062       0.0481      0.8263
        region    1     1     -0.6684    0.5034       1.7626      0.1843
        region    2     1      0.4737    0.4568       1.0754      0.2997
        region    3     0      0            .            .           .


                        Odds Ratio Estimates

                               Point          95% Wald
        Effect              Estimate     Confidence Limits

        lyears                1.316      0.996      1.740
        manner                0.266      0.126      0.560
        age                   1.028      1.000      1.057
        start                 1.041      0.999      1.086
        military              1.072      0.495      2.325
```

(*continued*)

Output 6.11 *(continued)*

conflict		0.784	0.401	1.534
loginc		1.060	0.649	1.731
literacy		1.034	1.016	1.052
region	0 vs 3	1.117	0.414	3.014
region	1 vs 3	0.513	0.191	1.375
region	2 vs 3	1.606	0.656	3.932

ALTERNATIVE APPROACHES TO MULTIPLE EVENT TYPES

Conditional Processes

The competing risks approach presumes that each event type has its own hazard model that governs both the occurrence and timing of events of that type. The appropriate imagery is one of independent causal mechanisms operating in parallel. Whichever type of event happens first, the individual is then no longer at risk of the other types.

This is clearly a defensible way of thinking about deaths due to natural causes, on the one hand, and forcible removal from power, on the other. It may not be so sensible, however, for the distinction between constitutional exits and nonconstitutional exits. We might imagine that a leader will stay in power as long as his popularity with key groups stays sufficiently high. When that popularity drops below a certain point, pressures will build for his removal. *How* he is removed is another question, and the answer depends on such things as the constitutional mechanisms that are available and the cultural traditions of the country. In this way of thinking about things, we have one mechanism governing the timing of events and another distinct mechanism determining the type of event, given that an event occurs.

For an even more blatant example, consider the event *buying a personal computer*, and suppose we subdivide this event into two types: buying a Macintosh computer and buying a Windows based computer. Now it would be absurd to suppose that there are two parallel processes here and that we merely observe whichever produces an event first. Rather, we have one process that governs the decision to buy a computer at all and another that governs which type is purchased. These kinds of situations arise most commonly when the different event types are alternative means for achieving some goal.

If we adopt this point of view, a natural way to proceed is to estimate one model for the timing of events (without distinguishing among event types) and a second model for the type of event (restricting the analysis to those individuals who experienced an event). For the timing of events, any of the models that we have considered so far are possible contenders. For the type of event, a binomial or multinomial logit model is a natural choice, although there are certainly alternatives. A major attraction of this approach is that there is no need to assume that the different kinds of events are uninformative for one another.

We have already estimated a logit model for constitutional versus nonconstitutional exits, with the results displayed in Output 6.11. There we interpreted the coefficients as differences in coefficients in the underlying hazard models. Now I am suggesting that we interpret these coefficients directly as determinants of whether a loss of power was by constitutional or nonconstitutional means, given that an exit from power occurred. We see, for example, that those who obtained power by nonconstitutional means are about four times as likely to lose power in the same way (based on the odds ratio of .266 for MANNER). On the other hand, each additional percentage point of literacy increases the odds that the exit will be constitutional rather than nonconstitutional by about 3 percent. Although the effects are only marginally significant, there are consistently positive effects of *time* on constitutional rather than nonconstitutional changes: older leaders, more recently installed leaders, and leaders who have been in power longer are all more likely to exit via constitutional mechanisms.

We still need a model for timing of exits from power, treating deaths from natural causes as censoring. Output 6.12 shows the results of estimating a Cox model, with the following statements:

```
PROC PHREG DATA=leaders;
   MODEL years*lost(0,2)=manner age start military conflict
      loginc literacy / TIES=EXACT;
   STRATA region;
RUN;
```

Because nearly half of the exits occurred at 0 or 1 year, I used the TIES=EXACT option.

Output 6.12 *Cox Model for Timing of Constitutional and Nonconstitutional Exits*

Parameter	DF	Parameter Estimate	Standard Error	Chi-Square	Pr > ChiSq	Hazard Ratio
manner	1	0.37642	0.15922	5.5892	0.0181	1.457
age	1	0.01718	0.00575	8.9381	0.0028	1.017
start	1	-0.01497	0.00840	3.1769	0.0747	0.985
military	1	-0.20906	0.16439	1.6174	0.2035	0.811
conflict	1	0.17099	0.13607	1.5792	0.2089	1.186
loginc	1	-0.24071	0.08860	7.3820	0.0066	0.786
literacy	1	0.00186	0.00331	0.3150	0.5747	1.002

Three results stand out in Output 6.12:

- Those who acquired power by nonconstitutional means had about a 46 percent higher risk of losing power (by means other than natural death).
- Each additional year of age increased the risk of exit by about 1.7 percent.
- A 1 percent increase in GNP per capita yielded about a .24 percent decrease in the risk of exit.

In my judgment, of all the analyses done in this chapter, Output 6.11 and Output 6.12 give the most meaningful representation of the processes governing the removal of leaders from power. To that, we may also wish to add Output 6.7, which shows that only age is associated with the risk of a death from natural causes.

Cumulative Incidence Functions

In equation (6.3) I defined the type-specific survivor function $S_j(t)$, but I warned that it did not apply to any well-defined, observable random variable. Nevertheless, if you do Kaplan-Meier (or life table) estimation with PROC LIFETEST, treating other events as censoring, what you get is an estimate of the function $S_j(t)$. This can be interpreted as the probability that an individual makes it to time t without having an event of type j, under the hypothetical presumption that other kinds of events cannot occur. Similarly, $F_j(t) = 1 - S_j(t)$ can be interpreted as the probability that an event of type j occurs to an individual by time t, *in the absence of any competing risks*.

In some fields, especially those that are more oriented toward applications, this is not regarded as a useful function to estimate, because the competing risks cannot be removed. A further reason is that

$$\sum_j F_j(t),$$

the sum of the probabilities of the different kinds of events occurring by time t, may often exceed 1.

Instead, many people prefer to estimate cumulative incidence functions, which are defined as

$$I_j(t) = \Pr(T_i < t, J_i = j).$$

This equation estimates the probability that an event of type j happens to individual i before time t, in the presence of competing event types. As described by Marubini and Valsecchi (1995), a consistent estimate of the cumulative incidence function is given by

$$\hat{I}_j(t) = \sum_{k|t_k \le t} \hat{S}(t_k) \frac{d_{jk}}{n_k}$$

where $\hat{S}(t)$ is the Kaplan-Meier estimate of the overall survivor function, d_{jk} is the number of events of type j that occurred at time t_k, and n_k is the number at risk at time t_k. The ratio of d_{jk} to n_k is an estimate of the hazard of event type j at time t_k.

An attractive feature of this estimator is that

$$\hat{S}(t) + \sum_j \hat{I}_j(t) = 1.$$

This says that at any time t, the probabilities for all the event types plus the probability of no event add up to 1. Another attraction is that the method does not depend on the assumption that each event type is non-informative for other event types.

Although cumulative incidence functions are not currently implemented in any SAS procedure, there is a macro called CUMINCID that is distributed with SAS and stored in its autocall location. That means that the macro is available for use in any SAS program. Here's how to use the macro for the LEADERS data set to estimate the cumulative incidence function for nonconstitutional exits (LOST=3):

```
%CUMINCID(DATA=leaders,
          TIME=years,
          STATUS=lost,
          EVENT=3,
          COMPETE=1 2,
          CENSORED=0)
```

In this version of the data set, leaders whose YEARS value was 0 were recoded to have a value of .5. Output 6.13 displays the results. Like the Kaplan-Meier method in PROC LIFETEST, the output produced by CUMINCID contains one line for every observation, but most of these lines contain no useful information. I've edited out all the lines that are not informative. The cumulative incidence probabilities are substantially lower than the cumulative failure probabilities reported by PROC LIFETEST using the Kaplan-Meier method. For example, at 24 years, the cumulative incidence estimate is .42 but the corresponding probability reported by LIFETEST is .63.

Output 6.13 *Estimates of the Cumulative Incidence Function*

				StdErr	Lower95Pct	Upper95Pct
Obs	years	censor	CumInc	CumInc	CumInc	CumInc
1	0.0000	.	0.00000	0.000000	.	.
54	0.5000	0	0.11229	0.014532	0.08713	0.14471
129	1.0000	0	0.15979	0.016919	0.12984	0.19664
185	2.0000	0	0.20689	0.018833	0.17309	0.24730
223	3.0000	0	0.23473	0.019810	0.19895	0.27696
262	4.0000	0	0.25637	0.020516	0.21916	0.29991
304	5.0000	0	0.28356	0.021330	0.24469	0.32860
337	6.0000	0	0.29932	0.021800	0.25951	0.34525
360	7.0000	0	0.31055	0.022151	0.27004	0.35715
374	8.0000	0	0.32554	0.022639	0.28406	0.37307
388	9.0000	0	0.33208	0.022874	0.29014	0.38008
399	10.0000	0	0.34618	0.023411	0.30320	0.39524
412	11.0000	0	0.36890	0.024258	0.32429	0.41964
427	13.0000	0	0.38142	0.024739	0.33588	0.43312
430	14.0000	0	0.38568	0.024896	0.33985	0.43770
437	15.0000	0	0.39467	0.025238	0.34818	0.44737
442	16.0000	0	0.39972	0.025468	0.35280	0.45289
445	17.0000	0	0.40513	0.025730	0.35771	0.45883
451	18.0000	0	0.41100	0.026042	0.36300	0.46534
455	19.0000	0	0.41749	0.026423	0.36878	0.47264

Cumulative Incidence Estimates with Confidence Limits Based on the Log Transform

The CUMINCID macro also has a STRATA parameter that produces cumulative incidence functions for multiple subgroups and plots the functions on a single graph. However, it does not have the capability of testing the null hypothesis that the cumulative incidence functions are identical, although several such tests have been proposed (for example, Gray, 1988). There are even regression methods based on cumulative incidence functions rather than on hazard functions (Fine and Gray, 1999; Scheike and Zhang, 2008). SAS macros that implement these methods can readily be found on the Web.

Despite the popularity of these methods (and the insistence by some that conventional methods are flatly incorrect), I think that cumulative incidence functions have been oversold. As a descriptive device in applied settings, the methods can certainly be useful. But if the aim is to understand the causal mechanisms underlying event occurrence, I still believe that the estimation of hazard functions (treating other events as censoring) is the way to go (Pintilie, 2006).

CONCLUSION

As we have seen, competing risks analysis is easily accomplished with conventional software by doing a separate analysis for each event type, treating other events as censoring. The biggest drawback of competing risks analysis is the requirement that times for different event types be independent or at least that each event be noninformative for the others. This requirement is exactly equivalent to the requirement for random censoring discussed in Chapter 2. In either case, violations can lead to biased coefficient estimates.

The seriousness of this problem depends greatly on the particular application. For the LEADERS data set, I argued that death due to natural causes is likely to be noninformative for the risk of either a constitutional or a nonconstitutional exit. For the latter two types, however, the presumption of noninformativeness may be unreasonable. In thinking about this issue, it is helpful to ask the question, "Are there unmeasured variables that may affect more than one event type?" If the answer is yes, then there should be a presumption of dependence.

Unfortunately, there's not a lot that can be done about the problem. It's possible to formulate models that incorporate dependence among event types but, for any such model, there's an independence model that does an equally good job of fitting the data. Dependence models typically impose parametric restrictions on the shape of the hazard functions, and the results may heavily depend on those restrictions. Heckman and Honoré

(1989) showed that you can identify nonparametric dependence models, so long as there is at least one continuous covariate with different coefficients for different event types (along with some other mild conditions), but the practical implications of their theorem have yet to be explored. Just because you can do something doesn't mean you can do it well.

So for all practical purposes, we have little choice but to use a method that rests on assumptions that may be implausible and cannot be tested. In Chapter 8, "Heterogeneity, Repeated Events, and Other Topics," I describe a sensitivity analysis for informative censoring that may also be useful for competing risks. Basically, this method amounts to redoing the analysis under two worst-case scenarios and hoping that the qualitative conclusions don't change. The other thing to remember is that you can reduce the problem of dependence by measuring and including those covariates that are likely to affect more than one type of event. For example, a diet that is high in saturated fat is thought to be a common risk factor for both heart disease and cancer. If a measure of dietary fat intake is included in the regression model, it can partially alleviate concerns about possible dependence among these two event types.

CHAPTER 7
Analysis of Tied or Discrete Data with PROC LOGISTIC

INTRODUCTION

This chapter shows you how to use the LOGISTIC procedure to analyze data in which many events occur at the same points in time. In Chapter 5, "Estimating Cox Regression Models with PROC PHREG," we looked at several different methods for handling tied data with PROC PHREG. There we saw that Breslow's method—the standard formula for partial likelihood estimation with tied data—is often a poor approximation when there are many ties. This problem was remedied by two exact methods, one that assumed that ties result from imprecise measurement and another that assumed that events really occur at the same (discrete) time. Unfortunately, both of these methods are computationally demanding for large data sets with many ties. We also looked at tied data in Chapter 4, "Estimating Parametric Regression Models with PROC LIFEREG," under the heading of interval censoring. While PROC LIFEREG is adept at estimating parametric models with interval censoring, it cannot incorporate time-dependent covariates.

The maximum likelihood methods described in this chapter do not suffer from these limitations. They do not rely on approximations, the computations are quite manageable even with large data sets, and they are particularly good at handling large numbers of time-dependent covariates. In addition, the methods make it easy to test hypotheses about the dependence of the hazard on time.

The basic idea is simple. Each individual's survival history is broken down into a set of discrete time units that are treated as distinct observations. After pooling these observations, the next step is to estimate a binary regression model predicting whether an event did or did not occur in each time unit. Covariates are allowed to vary over time from one time unit to another.

This general approach has two versions, depending on the form of the binary regression model. By specifying a logit link, you get estimates of the discrete-time proportional odds model proposed by Cox. This model is identical to the model estimated when you specify the TIES=DISCRETE option in PROC PHREG. Alternatively, by specifying a complementary log-log link, you get estimates of an underlying proportional hazards model in continuous time. This is identical to the model that is estimated when you specify the TIES=EXACT option in PROC PHREG.

The mechanics of this approach are similar to those of the piecewise exponential model described in Chapter 4 and the counting process syntax of Chapter 5. The main difference is that those methods assumed that you know the exact time of the event within a given interval. By contrast, the procedures in this chapter presume that you know only that an event occurred within a given interval.

THE LOGIT MODEL FOR DISCRETE TIME

We begin with the logit version of the model because it is more widely used and because logistic regression is already familiar to many readers. In Chapter 5 (see **The DISCRETE Method**), we considered Cox's model for discrete-time data. In brief, we let P_{it} be the conditional probability that individual i has an event at time t, given that an event has not already occurred to that individual. The model says that P_{it} is related to the covariates by a logistic regression equation:

$$\log\left[\frac{P_{it}}{1 - P_{it}}\right] = \alpha_t + \beta_1 x_{it1} + \ldots + \beta_k x_{itk} \tag{7.1}$$

where $t = 1, 2, 3, \ldots$. This model is most attractive when events can only occur at regular, discrete points in time, but it has also been frequently employed when ties arise from grouping continuous-time data into intervals.

In Chapter 5, we saw how to estimate this model by the method of partial likelihood, thereby discarding any information about the α_ts. Now we are going to estimate the same model by *maximum* likelihood, so that we get explicit estimates of the α_ts. The procedure is best explained by way of an example. As in Chapter 5, we'll estimate the model for 100 simulated job durations, measured from the year of entry into the job until the year that the employee quit. Durations after the fifth year are censored. We know only the year in which the employee quit, so the survival times have values of 1, 2, 3, 4, or 5. These values are contained in a variable called DUR, while the variable EVENT is coded 1 if the employee quit; otherwise, it is coded 0. Covariates are ED (years of education), PRESTIGE (a measure of the prestige of the occupation), and SALARY in the first year of the job. None of these covariates are time-dependent.

The first task is to take the original data set (JOBDUR) with one record per person and create a new data set (JOBYRS) with one record for each year that each person was observed. Thus, someone who quit in the third year gets three observations, while someone who still had not quit after five years on the job gets five observations. The following DATA step accomplishes this task:

```
DATA jobyrs;
   SET jobdur;
   DO year=1 TO dur;
      IF year=dur AND event=1 THEN quit=1;
      ELSE quit=0;
      OUTPUT;
   END;
RUN;
```

The DO loop creates 272 person-years that are written to the output data set. The IF statement defines the dependent variable QUIT, which equals 1 if the employee quit in that particular person-year; otherwise, QUIT equals 0. Thus, if a person quit in the fifth year, QUIT is coded 0 for the first four records and 1 in the last record. For people who don't quit during any of the 5 years, QUIT is coded 0 for all five records.

Output 7.1 shows the first 20 records produced by this DATA step. Observation 1 is for a person who quit in the first year of the job. Observations 2 through 5 correspond to a person who quit in the fourth year. QUIT is coded 0 for the first three years and 1 for the fourth. Observations 6 through 10 correspond to a person who still held the job at the end of the fifth year.

Output 7.1 *First 20 Cases of Person-Year Data Set for Job Durations*

Obs	dur	event	ed	prestige	salary	year	quit
1	1	1	7	3	19	1	1
2	4	1	14	62	17	1	0
3	4	1	14	62	17	2	0
4	4	1	14	62	17	3	0
5	4	1	14	62	17	4	1
6	5	0	16	70	18	1	0
7	5	0	16	70	18	2	0
8	5	0	16	70	18	3	0
9	5	0	16	70	18	4	0
10	5	0	16	70	18	5	0
11	2	1	12	43	135	1	0
12	2	1	12	43	135	2	1
13	3	1	9	18	12	1	0
14	3	1	9	18	12	2	0
15	3	1	9	18	12	3	1
16	1	1	11	31	12	1	1
17	1	1	13	26	6	1	1
18	1	1	10	1	4	1	1
19	2	1	12	28	17	1	0
20	2	1	12	28	17	2	1

Now we're ready to estimate a logistic regression model for these data. The following program accomplishes this task:

```
PROC LOGISTIC DATA=jobyrs;
   CLASS year / PARAM=GLM;
   MODEL quit(DESC)=ed prestige salary year;
RUN;
```

By specifying YEAR as a CLASS variable, we tell PROC LOGISTIC to treat this variable as categorical rather than quantitative. The PARAM=GLM option overrides the default effect coding and tells PROC LOGISTIC to create a set of four indicator (dummy) variables, with the reference category being YEAR=5 (the highest value). The DESC option specifies that the model will predict the probability of a 1 for QUIT rather than a 0.

Output 7.2 shows the results. Comparing these estimates with the partial likelihood estimates in Output 5.10, we see that the coefficients of ED, PRESTIGE, and SALARY are similar, as are the chi-square statistics. Again, this is not surprising because they are simply alternative ways of estimating the same model.

Output 7.2 *Maximum Likelihood Estimates of Discrete-Time Logistic Model for Job Duration Data*

```
                    Type 3 Analysis of Effects

                               Wald
          Effect        DF   Chi-Square    Pr > ChiSq

          ed            1       6.8392       0.0089
          prestige      1      46.5794       <.0001
          salary        1       6.6791       0.0098
          year          4      23.2529       0.0001



          Analysis of Maximum Likelihood Estimates

                              Standard       Wald
Parameter        DF  Estimate   Error   Chi-Square   Pr > ChiSq

Intercept        1    3.4443   1.1821     8.4896       0.0036
ed               1    0.2249   0.0860     6.8392       0.0089
prestige         1   -0.1235   0.0181    46.5794       <.0001
salary           1   -0.0268   0.0104     6.6791       0.0098
year     1       1   -2.6875   0.8327    10.4174       0.0012
year     2       1   -1.4475   0.7671     3.5605       0.0592
year     3       1   -0.0130   0.7272     0.0003       0.9857
year     4       1    0.2355   0.7779     0.0916       0.7621
year     5       0    0          0          .            .
```

Unlike partial likelihood, the maximum likelihood method also gives us estimates for the effect of time on the odds of quitting, as reflected in the α_ts in equation (7.1). The intercept in Output 7.2 is an estimate of α_5, the log-odds of quitting in year 5 for a person with values of 0 on all covariates. For level j of the YEAR variable, the coefficient is an estimate of $\alpha_j - \alpha_5$ (that is, the difference in the log-odds of quitting in year j and the log-odds of quitting in year 5, controlling for the covariates). We see that the log-odds is lowest in the first year of the job, rises steadily to year 3, and then stays approximately constant for the next 2 years. Overall, the effect of YEAR is highly significant with a Wald chi-square statistic of 23.25 with 4 d.f., as shown in the Type 3 table.

When the model in equation (7.1) is estimated by partial likelihood, there can be no restrictions on the α_ts. With the maximum likelihood method of this chapter, however, we can readily estimate restricted versions of the model. In fact, because time (in this case, YEAR) is just another variable in the regression model, we can specify the dependence of the hazard on time as any function that SAS allows in the DATA statement. For example, if we remove YEAR from the CLASS statement but keep it in the MODEL statement, we constrain the effect of YEAR to be linear on the log-odds of quitting. Alternatively, we can take the logarithm of YEAR before putting it in the model, or we can fit a quadratic model with YEAR and YEAR squared. To compare the fit of these models, we can use the AIC (Akaike's information criterion) or SC (Schwarz's criterion) statistics reported by PROC LOGISTIC. These are simply adjustments to −2 times the log-likelihood that penalize models for having more parameters. Here are the results:

	AIC	SC
Unrestricted	215.67	244.21
Linear	216.28	234.31
Quadratic	212.19	233.82
Logarithmic	211.98	230.00

For both fit measures, smaller values are better. Clearly, the logarithmic model is the best. The coefficients for ED, PRESTIGE, and SALARY in the logarithmic model (not shown) hardly change at all from the unrestricted model. The coefficient for the logarithm of YEAR is 2.00, indicating that a 1 percent increase in time in the job is associated with a 2 percent increase in the odds of quitting.

THE COMPLEMENTARY LOG-LOG MODEL FOR CONTINUOUS-TIME PROCESSES

As already noted, the logit model presumes that events can occur only at discrete points in time. For most applications, however, ties occur because event times are measured coarsely even though events can actually occur at any point in time. Aside from the implausibility of the logit model for such data, the logit model suffers from a lack of invariance to the length of the time interval. In other words, switching from person-months to person-years changes the *model* in a fundamental way, so that coefficients are not directly comparable across intervals of different length.

To avoid these difficulties, we can first specify a model for continuous-time data and from that derive a model for data grouped into intervals. (In essence, that's how the EXACT method was developed for PROC PHREG.) Suppose that the intervals are of equal length beginning at the origin. We'll index them by $t=1, 2, 3, \ldots$. As before, let P_{it} be the probability that an event occurs to individual i in interval t, given that the individual did not have events in any of the earlier intervals. If we now assume that events are generated by Cox's proportional hazards model, it follows (Prentice and Gloeckler, 1978) that

$$\log\left[-\log(1-P_{it})\right] = \alpha_t + \beta_1 x_{it1} + \ldots + \beta_k x_{itk} . \tag{7.2}$$

This result is easily obtained using equation (5.14) in Chapter 5. The transformation on the left side is called the *complementary log-log function*, which is also what we call the model.

Like the logit function, the complementary log-log function takes a quantity that varies between 0 and 1 and changes it to something that varies between minus and plus infinity. Unlike the logit function, however, the complementary log-log function is *asymmetrical*. For example, after taking the logit transformation, a change in probability from .25 to .50 is the same as from .50 to .75. On the complementary log-log scale, however, the difference between probabilities of .25 and .50 is larger than the difference between .50 and .75. This difference has an important practical implication. In the logit model, switching the values of the binary dependent variable merely changes the signs of the coefficients. In the complementary log-log model, on the other hand, switching the values produces completely different coefficient estimates, and it can even result in nonconvergence of the maximum likelihood algorithm. It's essential, then, that the model be set up to predict the probability of an *event* rather than a non-event.

Another important point about the model in equation (7.2) is that the β coefficients are identical to the coefficients in the underlying proportional hazards model. That doesn't mean that the estimates that you get from grouped data will be the same as those from the original ungrouped data. What it does mean is that both are estimating the same underlying parameters and are directly comparable to each other. It also means that the complementary log-log coefficients have a relative hazard interpretation, just like Cox model coefficients, and that the *model* (not the estimates) is invariant to interval length.

Originally, people preferred the logit model because there was little software for the complementary log-log model. Now there's no excuse. Many popular packages have complementary log-log options, and SAS makes it available in the LOGISTIC, PROBIT, GENMOD, and GLIMMIX procedures. In PROC LOGISTIC, PROC GENMOD, and PROC GLIMMIX, you specify it with the LINK=CLOGLOG option in the MODEL statement. In PROC PROBIT, you specify D=GOMPERTZ as an option in the MODEL statement. (D=GOMPERTZ is a misnomer. The correct name of the distribution function corresponding to the complementary log-log is *Gumbel*, not Gompertz.)

To estimate the complementary log-log model for the job duration data, we can use exactly the same SAS statements that we used for the logit model, except for the addition of the LINK=CLOGLOG to the MODEL statement. The coefficients in Output 7.3 are directly comparable to those in Output 5.9 for the EXACT method in PROC PHREG. Indeed, they are very similar—much closer to each other than either is to the two approximate methods in Output 5.9.

Output 7.3 *Maximum Likelihood Estimates of the Complementary Log-Log Model for Job Duration Data*

Parameter		DF	Estimate	Standard Error	Wald Chi-Square	Pr > ChiSq
Intercept		1	2.1952	0.8868	6.1270	0.0133
ed		1	0.1655	0.0632	6.8470	0.0089
prestige		1	-0.0926	0.0126	53.8990	<.0001
salary		1	-0.0229	0.00895	6.5623	0.0104
year	1	1	-2.0952	0.6600	10.0777	0.0015
year	2	1	-1.1097	0.6156	3.2489	0.0715
year	3	1	0.0489	0.5840	0.0070	0.9333
year	4	1	0.2131	0.6287	0.1149	0.7346
year	5	0	0	.	.	.

We can interpret the coefficients just as if this were a proportional hazards model. In particular, exponentiating the coefficients gives us hazard ratios. For example, a 1-year increase in education produces a $100(\exp(.1655)-1)=18$ percent increase in the hazard of quitting. A one-unit increase in prestige yields a $100(\exp(-.0926)-1) = -9$ percent change in the hazard of quitting. For a thousand-dollar increase in salary, we get a $100(\exp(-0.0229)-1) = -2$ percent change in the hazard of quitting.

Again, we see a strong effect of YEAR, with approximately the same pattern as in the logit model. As with that model, we can easily estimate constrained effects of YEAR by including the appropriate transformed variable as a covariate. Some of these constrained models have familiar names. If the logarithm of YEAR is a covariate, then we are estimating a Weibull model. If we just include YEAR itself (without the CLASS statement), we have the Gompertz model described briefly in Chapter 2, "Basic Concepts of Survival Analysis," but not otherwise available in SAS.

Comparing the results in Output 7.3 with those for the logit model in Output 7.2, we see that the coefficients are somewhat larger for the logit model, but the chi-square statistics and *p*-values are similar. This is the usual pattern. Only rarely do the two methods lead to different qualitative conclusions.

DATA WITH TIME-DEPENDENT COVARIATES

Because the maximum likelihood method is particularly effective at handling time-dependent covariates, let's look at an example with three covariates that change over time. The sample consists of 301 male biochemists who received their doctorates in 1956 or 1963. At some point in their careers, all of these biochemists had jobs as assistant professors at graduate departments in the U.S. The event of interest is a promotion to associate professor. The biochemists were followed for a maximum of 10 years after beginning their assistant professorships. For a complete description of the data and its sources, see Long, Allison, and McGinnis (1993). That article focuses on comparisons of men and women, but here I look only at men in order to simplify the analysis. The original data set has one record per person and includes the following variables:

DUR — the number of years from beginning of the job to promotion or censoring.

EVENT — has a value of 1 if the person was promoted; otherwise, EVENT has a value of 0.

UNDGRAD — a measure of the *selectivity* of undergraduate institution (ranges from 1 to 7).

PHDMED — has a value of 1 if the person received his Ph.D. from a medical school; otherwise, PHDMED has a value of 0.

PHDPREST — a measure of the prestige of the person's Ph.D. institution (ranges from 0.92 to 4.62).

ART1-ART10 — the cumulative number of articles the person published in each of the 10 years.

CIT1-CIT10 — the number of citations in each of the 10 years to all previous articles.

PREST1 — a measure of the prestige of the person's first employing institution (ranges from 0.65 to 4.6).

PREST2 — a measure of the prestige of the person's second employing institution (coded as missing for those who did not change employers). No one had more than two employers during the period of observation.

JOBTIME — the year of employer change, measured from the start of the assistant professorship (coded as missing for those who did not change employers).

The covariates describing the biochemists' graduate and undergraduate institutions are constant over time, but article counts, citation counts, and employer prestige all vary with time. The citation counts, taken from *Science Citation Index* (Institute for Scientific Information), are sometimes interpreted as a measure of the *quality* of a scientist's published work, but it may be more appropriate to regard them as a measure of the *impact* on the work of other scientists.

The first and most complicated step is to convert the file of 301 persons into a file of person-years. The following DATA step accomplishes that task:

```
DATA rankyrs;
   INFILE 'c:rank.dat';
   INPUT dur event undgrad phdmed phdprest art1-art10
         cit1-cit10 prest1 prest2 jobtime;
   ARRAY arts(*) art1-art10;
   ARRAY cits(*) cit1-cit10;
   IF jobtime=. THEN jobtime=11;
   DO year=1 TO dur;
      IF year=dur THEN promo=event;
         ELSE promo=0;
      IF year GE jobtime THEN prestige=prest2;
         ELSE prestige=prest1;
      articles=arts(year);
      citation=cits(year);
      OUTPUT;
   END;
RUN;
```

This DATA step has the same basic structure as the one for the job duration data. The DO loop creates and outputs a record for each person-year, for a total of 1,741 person-years. Within the DO loop, the dependent variable (PROMO) is assigned a value of 1 if a promotion occurred in that person-year; otherwise, PROMO has a value of 0.

What's new is that the multiple values of each time-dependent covariate must be read into a single variable for each of the person-years. For the article and citation variables, which change every year, we create arrays that enable us to refer to, say, articles in year 3 as ARTS(3). For the two job prestige variables, we must test to see whether the current year (in the DO loop) is greater than or equal to the year in which a change occurred. If it is, we assign the later value to the variable PRESTIGE; otherwise, PRESTIGE is assigned the earlier value. Note that for this to work, we recode the JOBTIME variable so that missing values (for people who didn't have a second employer) are recoded as 11. That way the DO loop index, which has a maximum value of 10, never equals or exceeds this value.

Now we can proceed to estimate regression models. For academic promotions, which usually take effect at the beginning of an academic year, it makes sense to view time as being truly discrete. A logit model, then, seems entirely appropriate. Using PROC LOGISTIC, we specify the model with a quadratic effect of YEAR:

```
PROC LOGISTIC DATA=rankyrs;
   MODEL promo(DESC)=undgrad phdmed phdprest articles citation
         prestige year year*year;
RUN;
```

Output 7.4 *Estimates of Logit Model for Academic Promotions*

Parameter	DF	Estimate	Standard Error	Wald Chi-Square	Pr > ChiSq
Intercept	1	-8.4845	0.7756	119.6539	<.0001
undgrad	1	0.1939	0.0635	9.3236	0.0023
phdmed	1	-0.2357	0.1718	1.8830	0.1700
phdprest	1	0.0271	0.0932	0.0843	0.7715
articles	1	0.0734	0.0181	16.3702	<.0001
citation	1	0.000126	0.00131	0.0091	0.9238
prestige	1	-0.2535	0.1138	4.9628	0.0259
year	1	2.0818	0.2338	79.3132	<.0001
year*year	1	-0.1586	0.0203	61.0097	<.0001

Odds Ratio Estimates

Effect	Point Estimate	95% Wald Confidence Limits	
undgrad	1.214	1.072	1.375
phdmed	0.790	0.564	1.106
phdprest	1.027	0.856	1.233
articles	1.076	1.039	1.115
citation	1.000	0.998	1.003
prestige	0.776	0.621	0.970

Output 7.4 shows the results. Not surprisingly, there is a strong effect of number of years as an assistant professor. The odds of a promotion increase rapidly with time, but at a decreasing rate; there is actually some evidence of a reversal after 7 years. I also estimated a model with a set of

10 dummy variables for years as an assistant professor, but a likelihood ratio chi-square test showed no significant difference between that model and the more restricted version shown here. Higher-order polynomials also failed to produce a significant improvement. On the other hand, the chi-square test for comparing the model in Output 7.4 and a model that excluded both YEAR and YEAR2 was 178.98 with 2 d.f.

Article counts had the next largest effect (as measured by the Wald chi-square test): each additional published article is associated with an increase of 7.6 percent in the odds of a promotion. But there is no evidence of any effect of citations, suggesting that it's the quantity of publications that matters in promotions, not the importance or impact of the published work.

Somewhat surprisingly, while there is no effect of the prestige of the institution where a biochemist got his doctorate, there is a substantial effect of the selectivity of his undergraduate institution. Each 1-point increase on the 7-point selectivity scale is associated with a 21 percent increase in the odds of a promotion, controlling for other covariates. There is also a slightly *negative* effect of the prestige of the current employer, suggesting that it may be harder to get promoted at a more prestigious department.

Notice that once the person-year data set is created, the time-dependent covariates are treated just like fixed covariates. Thus, many models can be estimated with the saved data set without additional data manipulations, making it especially convenient to estimate models with large numbers of time-dependent covariates.

ISSUES AND EXTENSIONS

In this section, we look at a number of complications and concerns that arise in the analysis of tied data using maximum likelihood methods.

Dependence Among the Observations?

A common reaction to the methods described in this chapter is that there must be something wrong. In general, when multiple observations are created for a single individual, it's reasonable to suppose that those observations are not independent, thereby violating a basic assumption used to construct the likelihood function. The consequence of dependence is usually biased standard error estimates and inflated test statistics. Even worse, there are different numbers of observations for different individuals, so some appear to get more weight than others.

While concern about dependence is often legitimate, it is not applicable here. In this case, the creation of multiple observations is not an ad-hoc method; rather, it follows directly from factoring the likelihood function for the data (Allison, 1982). The basic idea is this: In its original form, the likelihood for data with no censoring can be written as a product of probabilities over all n observations, as follows:

$$\prod_{j=1}^{n} \Pr(T_i = t_i) \tag{7.3}$$

where T_i is the random variable and t_i is the particular value observed for individual i. Each of the probabilities in equation (7.3) can be factored in the following way. If $t_i = 5$, we have

$$\Pr(T_i = 5) = P_{i5}(1 - P_{i4})(1 - P_{i3})(1 - P_{i2})(1 - P_{i1}) \tag{7.4}$$

where, again, P_{it} is the conditional probability of an event at time t, given that an event has not already occurred. This factorization follows directly from the definition of conditional probability. Each of the five terms in equation (7.4) behaves as if it came from a distinct, independent observation.

For those who may still be unconvinced, a comparison of the standard errors in Output 5.10 with those in Output 7.2 should be reassuring. They are virtually identical, despite the fact that the partial likelihood estimates in Output 5.10 are based on 100 persons, while the maximum likelihood estimates in Output 7.2 are based on 272 person-years.

This lack of dependence holds only when no individual has more than one event. When events are repeatable, as discussed in Chapter 8, "Heterogeneity, Repeated Events, and Other Topics," there is a real problem of dependence. But the problem is neither more nor less serious than it is for other methods of survival analysis.

Handling Large Numbers of Observations

Although the creation of multiple observations for each individual does not violate any assumptions about independence, it may cause practical problems when the number of individuals is large and the time intervals are small. For example, if you have a sample of 1,000 persons observed at monthly intervals over a 5-year period, you could end up with a working data set of nearly 60,000 person-months. While this is not an impossibly large number, it can certainly increase the time you spend waiting for results, thereby inhibiting exploratory data analysis.

If you find yourself in this situation, there are several options you may want to consider:

■ First, you should ask yourself if you really need to work with such small time intervals. If the time-dependent covariates are only changing annually, you might as well switch from person-months to person-years. True, there will be some loss of precision in the estimates, but this is usually minimal. And by switching to the piecewise exponential model described in Chapter 4 (which has a similar data structure), you can even avoid the loss of precision. On the other hand, if the time-dependent covariates *are* changing at monthly intervals, you should not aggregate to larger intervals.

■ Second, if your covariates are all categorical or they at least have a small number of levels, you can achieve great computational economy by estimating the models from data that are grouped by covariate values. You can accomplish this by using PROC SUMMARY to create a grouped data set and then using the grouped data syntax in PROC LOGISTIC. Alternatively, PROC CATMOD will automatically group the data when you specify a logit model, and you can save the grouped data set for further analysis.

■ Third, if you are estimating logit models, you may want to sample on the dependent variable, at least for exploratory analysis. Typically, the data sets created for the methods in this chapter have a dichotomous dependent variable with an extreme split—the number of nonevents will be many times larger than the number of events. What you can do is take *all* the observations with events and a random subsample of the observations without events, so that the two groups are approximately equal. Is this legitimate? Well, it is for the logit model (but not for the complementary log-log model). It is well known that disproportionate stratified sampling on the dependent variable in a logit analysis does not bias coefficient estimates (Prentice and Pike, 1979).

Unequal Intervals

To this point, we have assumed that the time intervals giving rise to tied survival times are all of equal length. It's not uncommon, however, for some intervals to be longer than others, either by design or by accident. For example, a study of deaths following surgery may have weekly follow-ups soon after the surgery when the risk of death is highest and then switch to monthly follow-ups later on. Some national panel studies, like the Panel Survey of Income Dynamics, were conducted annually in most years, but not in all. As a result, some intervals are 2 years long instead of

1. Even if the intervals are equal by design, it often happens that some individuals cannot be reached on some follow-ups.

Regardless of the reason, it should be obvious that, other things being equal, the probability of an event will increase with the length of the interval. And if time-dependent covariates are associated with interval length, the result can be biased coefficient estimates. The solution to this problem depends on the pattern of unequal intervals and the model being estimated.

There is one case in which no special treatment is needed. If you are estimating the models in equation (7.1) or equation (7.2), which place no restrictions on the effect of time, *and* if the data are structured so that every individual's interval at time *t* is the same length as every other individual's interval at time *t*, then the separate parameters that are estimated for every time interval automatically adjust for differences in interval length. This situation is not as common as you might think, however. Even when intervals at the same *calendar* time have the same length, intervals at the same *event* time will have different lengths whenever individuals have different origin points.

In all other cases, an ad-hoc solution will usually suffice: simply include the length of the interval as a covariate in the model. If there are only two distinct interval lengths, a single dummy variable will work. If there are a small number of distinct lengths, construct a set of dummy variables. If there are many different lengths, you will probably need to treat length as a continuous variable but include a squared term in the model to adjust for nonlinearity.

Empty Intervals

In some data sets, there are time intervals in which no individual experiences an event. For example, in the LEADER data set that we analyzed in Chapter 6, "Competing Risks," none of the 472 leaders lost power in the 22nd year of rule. Naturally, this is most likely to occur when the number of time intervals is large and the number of individuals is small. Whenever there are *empty* intervals, if you try to estimate a model with an unrestricted effect of time, as in equation (7.1) or equation (7.2), the maximum likelihood algorithm will not converge. This result is a consequence of the following general principle. For any dichotomous covariate, consider the 2×2 contingency table formed by that covariate and the dichotomous dependent variable. If any of the four cells in that table has a frequency of 0, the result is nonconvergence.

An easy solution is to fit a model with restrictions on the time effect. The quadratic function that was estimated in Output 7.4, for example, will not suffer from this problem. Alternatively, if you don't want to lose the flexibility of the unrestricted model, you can constrain the coefficient for any empty interval to be the same as that of an adjacent interval. The simplest way to do this is to recode the variable containing the interval values so that the adjacent intervals have the same value. For the LEADER data, this requires a DATA step with the following statement:

```
IF year=22 THEN year=21;
```

Then specify YEAR as a CLASS variable in PROC LOGISTIC. Instead of separate dummy (indicator) variables for years 21 and 22, this code produces one dummy variable equal to 1 if an event time was equal to either of those 2 years; otherwise, the variable is equal to 0.

Left Truncation

In Chapter 5, we discussed a problem known as left truncation, in which individuals are not at risk of an event until some time after the origin time. This commonly occurs when "survivors" are recruited into a study at varying points in time. We saw how this problem can be easily corrected with the partial likelihood method using PROC PHREG. The solution is equally easy for the maximum likelihood methods discussed in this chapter, although quite different in form. In creating the multiple observations for each individual, you simply delete any time units in which the individual is known not to be at risk of an event. For example, if patients are recruited into a study at various times since diagnosis, no observational units are created for time intervals that occurred before recruitment. We can still include time since diagnosis as a covariate, however.

This method also works for temporary withdrawals from the risk set. If your goal is to predict migration of people from Mexico to the U.S., you will probably want to remove anyone from the risk set when he or she was in prison, in the military, and so on. Again, you simply exclude any time units in which the individual is definitely not at risk.

Competing Risks

In Chapter 6, we saw that the likelihood function for data arising from multiple event types can be factored into a separate likelihood function for each event type, treating other event types as censoring. Strictly speaking, this result only holds when time is continuous and measured precisely, so that there are no ties. When time is discrete or when continuous-time data are grouped into intervals, the likelihood function does *not* factor. If you want full-information maximum likelihood estimates, you must estimate a model for all events simultaneously. On the other hand, it's also possible to do separate analyses for each event type without biasing the parameter estimates and with only slight loss of precision.

These points are most readily explained for the logit model for competing risks. Let P_{ijt} be the conditional probability that an event of type j occurs to person i at time t, given that no event occurs before time t. The natural extension of the logit model to multiple event types is the multinomial logit model:

$$P_{ijt} = \frac{\exp(\boldsymbol{\beta}_j \mathbf{x}_{it})}{1 + \sum_k \exp(\boldsymbol{\beta}_k \mathbf{x}_{it})}, \qquad j=1, 2, \ldots$$

or, equivalently

$$\log\left(\frac{P_{ijt}}{P_{i0t}}\right) = \boldsymbol{\beta}_j \mathbf{x}_{it} \qquad j=1, 2, \ldots$$

where P_{i0t} is the probability that no event occurs at time t to individual i. As in the case of a single event type, the likelihood for data arising from this model can be manipulated so that each discrete-time point for each individual appears as a separate observation. If there are, say, three event types, these individual time units can have a dependent variable coded 1, 2, or 3 if an event occurred and coded 0 if no event occurred. We can then use PROC LOGISTIC to estimate the model simultaneously for all event types.

For the job duration data analyzed earlier in this chapter, suppose that there are actually two event types, quitting (EVENT=1) and being fired (EVENT=2), with EVENT=0 for censored cases. The following statements produce a person-year data set with a dependent variable called OUTCOME, which is coded 1 for quit, 2 for fired, and 0 for neither:

```
DATA jobyrs2;
   SET jobdur;
   DO year=1 TO dur;
     IF year=dur THEN outcome=event;
     ELSE outcome=0;
     OUTPUT;
   END;
RUN;
```

We can then use PROC LOGISTIC to estimate the multinomial logit model:

```
PROC LOGISTIC data=jobyrs2;
  MODEL outcome(REF='0')=ed prestige salary year
     / LINK=GLOGIT;
RUN;
```

The REF='0' option makes 0 the reference category for the dependent variable. The default is to use the highest value as the reference category, but in survival analysis we want each event type to be compared with no event. The LINK=GLOGIT option requests an unordered multinomial logit model, rather than the default cumulative logit model.

Output 7.5 displays the results. The Type 3 table gives statistics for testing the null hypothesis that *both* coefficients for each covariate are 0, a hypothesis that is clearly rejected for ED, PRESTIGE, and YEAR. The effect of SALARY is less dramatic but still significant at the .03 level. The lower portion of the table gives the coefficient estimates and their respective test statistics. All the parameters labeled 1 pertain to the contrast between type 1 (quit) and no event, while all the parameters labeled 2 pertain to the contrast between type 2 (fired) and no event. We see that education increases the odds of quitting but reduces the odds of being fired, with both coefficients highly significant. The prestige of the job, on the other hand, reduces the risk of quitting while increasing the risk of being fired. Again, both coefficients are highly significant. Finally, the odds of quitting increase markedly with each year, but the odds of being fired hardly change at all.

Output 7.5 *PROC LOGISTIC Results for Competing Risks Analysis of Job Data*

```
                  Type 3 Analysis of Effects

                                 Wald
           Effect      DF    Chi-Square    Pr > ChiSq

           ed           2       21.0701       <.0001
           prestige     2       66.5608       <.0001
           salary       2        6.7535       0.0342
           year         2       19.5623       <.0001
```

(*continued*)

Output 7.5 *(continued)*

```
               Analysis of Maximum Likelihood Estimates

                                      Standard       Wald
       Parameter  outcome  DF  Estimate    Error  Chi-Square  Pr > ChiSq

       Intercept  1         1    0.3286   0.8196      0.1608      0.6885
       Intercept  2         1   -1.5491   1.6921      0.8381      0.3599
       ed         1         1    0.1921   0.0836      5.2809      0.0216
       ed         2         1   -0.7941   0.2011     15.5874     <.0001
       prestige   1         1   -0.1128   0.0169     44.4661     <.0001
       prestige   2         1    0.1311   0.0274     22.8174     <.0001
       salary     1         1   -0.0255   0.0103      6.1743      0.0130
       salary     2         1    0.0110   0.0150      0.5333      0.4652
       year       1         1    0.7725   0.1748     19.5367     <.0001
       year       2         1   -0.00962  0.2976      0.0010      0.9742
```

Now, we'll redo the analysis with separate runs for each event type. For this, we rely on a well-known result from multinomial logit analysis (Begg and Gray, 1984). To estimate a model for event type 1, simply eliminate from the sample all the person-years in which events of type 2 occurred. Then, do a binomial logit analysis for type 1 versus no event. To estimate a model for event type 2, eliminate all the person-years in which events of type 1 occurred. Then, do a binomial logit analysis for type 2 versus no event. Here's the SAS code that accomplishes these tasks:

```
PROC LOGISTIC DATA=jobyrs2;
   WHERE outcome NE 2;
   MODEL outcome(DESC)=ed prestige salary year;
RUN;

PROC LOGISTIC DATA=jobyrs2;
   WHERE outcome NE 1;
   MODEL outcome(DESC)=ed prestige salary year;
RUN;
```

This procedure is justified as a form of conditional maximum likelihood. The resulting estimates are consistent and asymptotically normal, but there is some loss of precision, at least in principle. In practice, both the coefficients and their estimated standard errors usually differ only trivially from those produced by the simultaneous estimation procedure. We can see this by comparing the results in Output 7.6 with those in Output 7.5. The advantages of separating the estimation process are that you can

- focus on only those event types in which you are interested
- specify quite different models for different event types, with different covariates and different functional forms.

Output 7.6 *PROC LOGISTIC Results for Separate Estimation of Quitting and Firing Models*

Quittings

Parameter	DF	Estimate	Standard Error	Wald Chi-Square	Pr > ChiSq
Intercept	1	0.3672	0.8254	0.1979	0.6564
ed	1	0.1895	0.0840	5.0911	0.0240
prestige	1	-0.1125	0.0169	44.3193	<.0001
salary	1	-0.0255	0.0103	6.1688	0.0130
year	1	0.7637	0.1742	19.2200	<.0001

Firings

Parameter	DF	Estimate	Standard Error	Wald Chi-Square	Pr > ChiSq
Intercept	1	-1.5463	1.7054	0.8221	0.3646
ed	1	-0.7871	0.1999	15.5022	<.0001
prestige	1	0.1300	0.0274	22.5041	<.0001
salary	1	0.0118	0.0155	0.5845	0.4445
year	1	-0.0296	0.2913	0.0103	0.9190

Now, what about competing risks for the complementary log-log model? Here, things are a little messier. It is possible to derive a multinomial model based on a continuous-time proportional hazards model, and this could be simultaneously estimated for all event types using maximum likelihood. This is not a standard problem, however, and no SAS procedure will do it without a major programming effort. Instead, we can use the same strategy for getting separate estimates for each event type that we just saw in the case of the logit model. That is, we delete all individual time units in which events other than the one of interest occurred. Then we estimate a dichotomous complementary log-log model for the event of interest versus no event. In effect, we are deliberately censoring the data at the beginning of any intervals in which other events occur. This should not be problematic because, even if we have continuous-time data, we have to assume that the different event types are noninformative for one another. Once we eliminate all extraneous events from the data, we reduce the problem to one for a single event type. Again, there will be some slight loss of information in doing this.

CONCLUSION

The maximum likelihood methods discussed in this chapter are attractive alternatives to partial likelihood when there are many ties and many time-dependent covariates. Not only are they much more computationally efficient, but they also give direct estimates of the effect of time on the hazard of an event. The methods discussed here have considerable intuitive appeal and are relatively straightforward to implement. Once the expanded data set is constructed, the analyst can proceed as in an ordinary logistic regression analysis with no need to treat time-dependent covariates any differently than fixed covariates. This approach also has advantages whenever there is ambiguity about the time origin. Because time is treated just like any other covariate, there is great flexibility in specifying and testing alternative functional forms, and multiple time scales with different origins can be included in the model.

256

CHAPTER **8**
Heterogeneity, Repeated Events, and Other Topics

INTRODUCTION

This chapter deals with several issues that arise for all the methods that we have previously discussed. The first two issues—heterogeneity and repeated events—are closely related. One problem with the models that we have been considering is that biological and social entities usually differ in ways that are not fully captured by the model. This *unobserved heterogeneity* can produce misleading estimates of hazard functions and attenuated estimates of covariate effects. When individuals can experience more than one event, unobserved heterogeneity can also produce dependence among the observations, leading to biased standard errors and test statistics. We'll survey a variety of methods for dealing with these problems. Later in the chapter, we'll see how to compute a generalized R^2 and how to gauge the possible consequences of informative censoring.

UNOBSERVED HETEROGENEITY

An implicit assumption of all the hazard models that we have considered so far is that if two individuals have identical values on the covariates, they also have identical hazard functions. If there are no covariates in the model, then the entire sample is presumed to have a single hazard function. Obviously, this is an unrealistic assumption. Individuals and their environments differ in so many respects that no set of measured covariates can possibly capture all the variation among them. In an ordinary linear regression model, this residual or unobserved heterogeneity is explicitly represented by a random disturbance term, for example,

$$y = \boldsymbol{\beta}\mathbf{x} + \varepsilon$$

where ε represents all unmeasured sources of variation in y. But in a Cox regression model, for example, there is no disturbance term:

$$\log h(t) = \alpha(t) + \boldsymbol{\beta}\mathbf{x} .$$

The absence of a random disturbance term does not mean that the model is deterministic. There is plenty of room for randomness in the relationship between the unobserved hazard $h(t)$ and the observed event time. Nevertheless, the presence of unobserved heterogeneity can lead to incorrect conclusions about the dependence of the hazard function on time, and it can also produce biased estimates of the covariate coefficients.

The most serious problem is this: unobserved heterogeneity tends to produce estimated hazard functions that decline with time, *even when the true hazard is not declining for any individual in the sample* (Proschan, 1963; Heckman and Singer, 1985). This fact is most easily explained by an example. Suppose we have a sample of 100 people, all of whom have hazards that are constant over time. The sample is equally divided between two kinds of people: those with a high hazard of death ($h(t)=2.0$) and those with a low hazard of death ($h(t)=0.5$). Unfortunately, we don't know which people have which hazard, so we must estimate a hazard function for the entire sample. Figure 8.1 shows what happens. The empirical hazard function starts out, as you might expect, midway between .5 and 2. But then it steadily declines until it approaches .5 as an asymptote. What's happening is that the high hazard people are dying more rapidly at all points in time. As a result, as time goes by, the remaining sample (the risk set) is increasingly made up of people with low hazards. Because we can estimate the hazard function at time t with only those who are still at risk at time t, the estimated hazard will be more and more like the lower hazard.

Figure 8.1 *Empirical Hazard Function Produced by Mixing Two Constant Hazards*

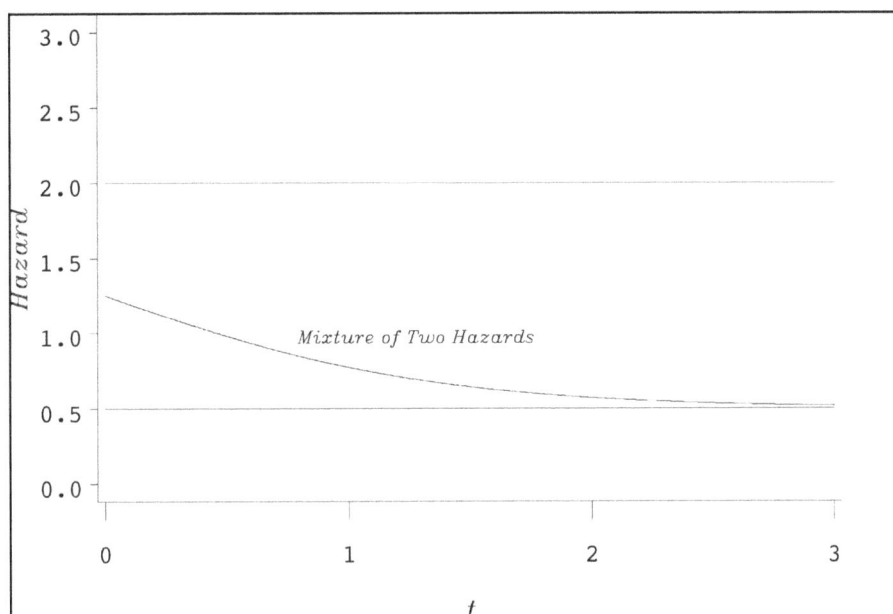

Of course, any real life situation will have more than two kinds of people, but the basic principle is the same. Those with higher hazards will tend to die (or experience whatever event is being studied) before those with lower hazards, leaving a risk set that is increasingly made up of low hazard people.

This problem has led to some potentially serious errors in interpreting research results. In the management literature, for example, there has been great interest in the hypothesis of the *liability of newness* (Hannan and Freeman, 1984). This hypothesis says that when firms are just starting out, they are prone to failure because they lack capital, contacts, traditions, and so on. As firms get older and accumulate resources, they become more resistant to failure. Empirical studies invariably show that, in fact, younger firms do have higher rates of failure than older firms. But this fact does not necessarily prove the hypothesis. The results are equally consistent with the hypothesis that firms differ in their initial vulnerability to failure. Weaker firms go out of business quickly while stronger firms survive.

What can be done? As we'll see in the next section, when events are repeatable, it is quite feasible to separate the true hazard function from unobserved heterogeneity. But when events are not repeatable, as with both human and organizational death, the options are limited. If you find an empirical hazard function that is *increasing*, then you can validly conclude that the true hazard is increasing for at least some fraction of the sample over some interval of time. But decreasing hazard functions are inherently ambiguous.

There have been numerous attempts to separate the hazard function from unobserved heterogeneity by formulating models that incorporate both. For example, we can insert a random disturbance term into a Weibull hazard model:

$$\log h(t) = \alpha \log t + \boldsymbol{\beta}\mathbf{x} + \varepsilon.$$

But models like this are highly sensitive to the choice of a distribution for ε or the form of the dependence on time. If such models are identified at all, it is only because of the imposition of a particular functional form. That situation is not conducive to drawing reliable conclusions.

What about estimates of the β coefficients? Are they affected by unobserved heterogeneity? Given what we know about linear models, it's a foregone conclusion that coefficients may be severely biased if the unobserved components are correlated with the measured covariates. The more interesting question is what happens when the unobserved disturbance is *independent* of the measured covariates. Early literature on this question seemed to suggest that coefficient estimates could be grossly biased in unexpected ways. But the work by Gail and colleagues (1984) is more persuasive. They showed that unobserved heterogeneity tends to attenuate the estimated coefficients toward 0. On the other hand, standard errors and test statistics are *not* biased. Therefore, a test of the hypothesis that a coefficient is 0 remains valid, even in the presence of unobserved heterogeneity. It's also important to realize that the attenuation of coefficients is not a problem unique to hazard models, but it occurs with a wide variety of nonlinear models, including logistic regression (Allison, 1987).

REPEATED EVENTS

All the models and analyses in the preceding chapters presume that no individual experiences more than one event. That's a reasonable presumption if the event is death. But most events in the social sciences are repeatable: births, marriages, job changes, promotions, arrests, residence changes, and so on. There are also many repeatable events that are of interest to biomedical scientists: tumor recurrences, seizures, urinary infections, and hospitalizations, to name only a few. Because so much of survival analysis has focused on deaths, good methods for handling repeated events were relatively slow in coming. Nevertheless, several important developments have occurred in recent years, and some of those methods are now available in SAS. Before examining the newer methods, however, we need to consider what's wrong with conventional methods.

Problems with Conventional Methods

There are basically two approaches to analyzing repeated events with standard software. First, you can do a separate analysis for each successive event. Suppose, for example, that you have reproductive histories for a sample of women who are married or who have ever been married, and you want to estimate a model for birth intervals. You start with an analysis for the interval between marriage and the first birth. For all those women who had a first birth, you then do a second analysis for the interval between first birth and second birth. You continue in this fashion until the number of women gets too small to reliably estimate a model.

For a more detailed example, let's look at some simulated data on job changes. For 100 people, I generated repeated job durations, from the point of entry into the labor market until 20 years had elapsed, at which time observation was terminated. Thus, people whose first job lasted more than 20 years were censored at year 20. Everyone else had exactly one censored job (the one still in progress in year 20) and at least one uncensored job. A total of 395 jobs were produced.

Each person was assigned a certain number of years of schooling (ED) that did not vary over jobs. Each job had a prestige score (ranging from 1 to 100), as well as a salary (in thousands of dollars) that remained constant during the job. The duration for each job was generated by a Weibull model that included ED, PRESTIGE, and the logarithm of SALARY as covariates. Coefficients were invariant across people and across time.

Output 8.1 displays the first 20 cases out of the 395 jobs in the working data set. ID is a variable that distinguishes different people. We see that these 20 jobs were held by four people: 6 jobs for person 1, 1 job for person 2, 3 for person 3, and 10 for person 4. EVENT is the censoring indicator. The variable SEQ simply keeps track of where each job is in a person's sequence of jobs. LOGSAL is the natural logarithm of salary.

Output 8.1   *First 20 Cases in Repeated Job Duration Data Set*

OBS	ID	EVENT	ED	SEQ	DURATION	PRESTIGE	LOGSAL
1	1	1	7	1	2.3575	27	2.30259
2	1	1	7	2	4.5454	31	1.38629
3	1	1	7	3	1.0864	8	2.07944
4	1	1	7	4	3.5893	22	2.83321
5	1	1	7	5	4.6676	30	3.40120
6	1	0	7	6	3.7538	24	3.13549
7	2	0	14	1	20.0000	69	3.33220
8	3	1	16	1	7.3753	69	3.58352
9	3	1	16	2	11.2632	60	3.85015
10	3	0	16	3	1.3615	56	3.82864
11	4	1	12	1	1.1436	27	2.56495
12	4	1	12	2	0.7880	35	2.19722
13	4	1	12	3	1.0437	30	2.77259
14	4	1	12	4	0.2346	15	2.56495
15	4	1	12	5	0.7983	39	3.21888
16	4	1	12	6	1.2168	41	3.52636
17	4	1	12	7	0.9160	24	3.43399
18	4	1	12	8	0.3174	26	3.36730
19	4	1	12	9	0.6573	45	3.82864
20	4	1	12	10	1.7036	41	3.91202

The easiest way to run separate models for each job in the sequence is to sort the data by SEQ and then include a BY statement in the PROC PHREG program:

```
PROC SORT DATA=jobmult;
   BY seq;
PROC PHREG DATA=jobmult;
   MODEL duration*event(0)=prestige logsal ed;
   BY seq;
RUN;
```

In Output 8.2, we see results for the first five job durations. As is typical of data like these, the number of observations declines substantially for each successive interval until, by the sixth job, it's really too small to support a credible analysis. The effect of PRESTIGE is fairly consistent over the five models, with more prestigious jobs having lower hazards of termination. The effect of ED is not significant in the second job, but it is clearly significant in the others. Its coefficient is much smaller for the first two jobs than for the later jobs. The salary effect varies greatly from job to job and hovers around the .05 level of significance, except for job 4, when it is far from significant.

Output 8.2 *PROC PHREG Coefficients for Successive Jobs (Selected Output)*

First Job (n=100)

Parameter	DF	Parameter Estimate	Standard Error	Chi-Square	Pr > ChiSq	Hazard Ratio
prestige	1	-0.08714	0.01212	51.6530	<.0001	0.917
logsal	1	-0.43249	0.21767	3.9478	0.0469	0.649
ed	1	0.17226	0.06597	6.8171	0.0090	1.188

Second Job (n=72)

Parameter	DF	Parameter Estimate	Standard Error	Chi-Square	Pr > ChiSq	Hazard Ratio
prestige	1	-0.07141	0.01555	21.0889	<.0001	0.931
logsal	1	-0.77067	0.28446	7.3399	0.0067	0.463
ed	1	0.09044	0.06747	1.7968	0.1801	1.095

Third Job (n=52)

Parameter	DF	Parameter Estimate	Standard Error	Chi-Square	Pr > ChiSq	Hazard Ratio
prestige	1	-0.11049	0.02365	21.8185	<.0001	0.895
logsal	1	-0.88703	0.41098	4.6583	0.0309	0.412
ed	1	0.43346	0.14648	8.7569	0.0031	1.543

Fourth Job (n=34)

Parameter	DF	Parameter Estimate	Standard Error	Chi-Square	Pr > ChiSq	Hazard Ratio
prestige	1	-0.10120	0.02613	14.9978	0.0001	0.904
logsal	1	-0.19497	0.44619	0.1909	0.6621	0.823
ed	1	0.43956	0.12108	13.1784	0.0003	1.552

Fifth Job (n=34)

Parameter	DF	Parameter Estimate	Standard Error	Chi-Square	Pr > ChiSq	Hazard Ratio
prestige	1	-0.07592	0.02744	7.6559	0.0057	0.927
logsal	1	-1.14765	0.58605	3.8349	0.0502	0.317
ed	1	0.34951	0.11327	9.5216	0.0020	1.418

While this approach has some things going for it, it's inefficient in a couple of respects. It's tedious to do multiple analyses, and the more numbers you have to interpret, the more room there is for ambiguity and confusion. While it is tempting to interpret the job-to-job variations as real differences, I know for a fact that the underlying process is constant over time—after all, I generated the data! If the process is, in fact, invariant from one interval to the next, it's also statistically inefficient to produce several redundant estimates. That may not be much of a concern if you're working with census data and have 100,000 cases. But if you have only 100 cases, you want to use your limited information to the best advantage. An additional problem is that later jobs are a biased sample: the only people who had a fifth job were those who already had four jobs in the 20-year period. Of necessity, those four jobs were shorter than average, so it's likely that their fifth job will also be short.

There is a second general approach to repeated events that avoids these problems by treating each interval as a distinct observation, pooling all the intervals together, and estimating a single model. Output 8.3 shows the results of doing that for the 395 job durations, also including SEQ as one of the predictors. We now see that all four variables have highly significant coefficients. This model is essentially equivalent to the gap-time model of Prentice and colleagues (1981), except that their model would have stratified on SEQ instead of including it as a covariate.

Output 8.3 *PROC PHREG Results for Pooled Job Durations*

Parameter	DF	Parameter Estimate	Standard Error	Chi-Square	Pr > ChiSq	Hazard Ratio
prestige	1	-0.07865	0.00633	154.3027	<.0001	0.924
logsal	1	-0.59694	0.11443	27.2127	<.0001	0.550
ed	1	0.18663	0.02887	41.7959	<.0001	1.205
seq	1	0.26911	0.02946	83.4457	<.0001	1.309

Unfortunately, this method introduces a new problem—dependence among the multiple observations. It's well-known that whenever two or more observations come from the same unit (person, litter, organization), they tend to be more alike than two randomly chosen observations. In the birth interval example, we would expect that women who have short first intervals will also tend to have short second intervals, and so on. Pooling these observations without taking the dependence into account can lead to standard error estimates that are biased downward and test statistics that are biased upward. In essence, the estimation procedure is fooled into thinking it has more information than it actually does. Because observations are correlated, some of the apparent information in the sample is redundant.

Dependence among observations can be thought of as arising from unobserved heterogeneity. Second intervals tend to be like first intervals because there are unmeasured, stable factors affecting both intervals. If we could measure all of these factors and include them as covariates, the dependence problem would disappear. But, of course, that's not going to happen. As we saw in the preceding section, unobserved heterogeneity leads to artifactually declining hazard functions and coefficients that are attenuated toward 0. These two problems are still present with repeated events, but with the added problem of biased standard errors and test statistics. That's important to remember because some methods correct *only* the standard errors, while others can also correct biases in the coefficients and hazard functions.

Before we look at ways of correcting for dependence, let's first look at a simple way to *detect* it. Even though dependence is always likely to be present to some degree, it may be so small that it has trivial effects on the estimates. Hence, we need some way to judge just how substantial the problem is. Here's a simple ad-hoc way to do just that: Estimate a model for the *second* interval with the length of the *first* interval as a covariate. You should also include the covariates you would otherwise put in the model because the important question is whether there is *residual* dependence after the effects of any covariates have been removed. Here is the code:

```
PROC SORT DATA=jobmult;
  BY id seq;
DATA joblag;
  SET jobmult;
  durlag=LAG1(duration);
PROC PHREG DATA=joblag;
  WHERE seq = 2;
  MODEL duration*event(0)=prestige logsal ed durlag;
RUN;
```

Output 8.4 shows the results for the 72 second jobs in the job duration data set. Two things are noteworthy here. First, the duration of the first job has a highly significant negative coefficient, indicating that long first jobs are associated with low hazards, implying long second jobs. Clearly, there is dependence here that needs to be corrected in some way. If the coefficient had *not* been significant, we could ignore the dependence without much fear of error. Second, the coefficients for the other covariates have changed substantially from the values in Output 8.2 for second jobs. While this is a further indication that dependence may be influencing the results in some way, these coefficients are not necessarily any better than those that we have already estimated, and they may even be worse. In other words, this technique is only good for diagnosing the problem, not for correcting it.

Output 8.4 *Model for Second Interval with First Interval as a Covariate*

Variable	DF	Parameter Estimate	Standard Error	Chi-Square	Pr > Chi-Square	Hazard Ratio
prestige	1	-0.060441	0.01576	14.71353	0.0001	0.941
logsal	1	-0.283335	0.32508	0.75966	0.3834	0.753
ed	1	0.036648	0.06953	0.27785	0.5981	1.037
dur1	1	-0.269221	0.07639	12.42143	0.0004	0.764

Given the dependence in the job durations, we can't accept the pooled estimates in Output 8.3 at face value. At a minimum, the standard errors are likely to be too low and the chi-square statistics too high. The coefficients may also be attenuated toward 0, and there may be other biases if the unobserved heterogeneity is correlated with the measured covariates. (Note that these coefficient biases will also occur in the unpooled estimates in Output 8.2.) What can we do about it? One approach is to ignore the possible biases and concentrate on getting better estimates of the standard errors and test statistics. The other approach is to formulate a model that incorporates unobserved heterogeneity and estimate that model by maximum likelihood or conditional likelihood. While this second approach can correct biases in the coefficients, it depends more on a correct specification of the model.

Robust Standard Errors

PROC PHREG has an option called COVSANDWICH that makes it easy to correct for dependence when there are repeated observations. This option invokes a method variously known as the *robust variance estimator* or the *modified sandwich estimator*, developed for Cox regression by Lin and Wei (1989) and described in some detail in Therneau and Grambsch (2000). (The name *sandwich* comes from the matrix formula for the robust variance-covariance estimator, which has the general form ABA where A is the inverse of the observed information matrix.) The technique is sometimes described as a *population-averaged* method. One attraction of this method is that there is no need to make assumptions about the nature or structure of the dependence. On the other hand, there is no correction for biases in the coefficients that arise from unobserved heterogeneity.

Here's a PROC PHREG program for the job duration data that includes this option:

```
PROC PHREG DATA=jobmult COVSANDWICH(AGGREGATE);
   MODEL duration*event(0)=prestige logsal ed seq;
   ID id;
RUN;
```

The option COVSANDWICH can be abbreviated to COVS. To correct for dependence, it's necessary to include both the AGGREGATE option and an ID statement that gives the name of the variable containing the ID number that has the same value for all observations in the same cluster (a person, in this example).

In Output 8.5, we see that the coefficients (and hazard ratios) are identical to what we got in Output 8.3 without the COVSANDWICH option. What has changed are the standard errors, chi-squares, and *p*-values. We also see a new column, "StdErr Ratio," that gives the ratio of the new, robust standard errors to the original standard errors. Three of the four standard errors are larger with the robust correction, which is what you typically find when you correct for dependence among repeated observations. However, the standard error for PRESTIGE is a little smaller. It's also common to find the greatest increase in the standard errors for those variables, like ED, that do not change over time.

Another change in the output is that there are now two versions of the score chi-square and the Wald chi-square for testing the null hypothesis that all the coefficients are 0. The model-based versions are what you get when you don't use the COVSANDWICH option. The sandwich chi-squares correct for dependence and will usually be smaller than the model-based chi-squares. In this example, the score statistic gets a much larger correction than the Wald statistic, although it's not obvious why this should be so.

Output 8.5 *PROC PHREG Results with Robust Standard Errors*

```
                 Testing Global Null Hypothesis: BETA=0

        Test                    Chi-Square       DF      Pr > ChiSq

        Likelihood Ratio          433.3143        4        <.0001
        Score (Model-Based)       386.7275        4        <.0001
        Score (Sandwich)           51.6611        4        <.0001
        Wald (Model-Based)        349.3395        4        <.0001
        Wald (Sandwich)           334.6795        4        <.0001


                 Analysis of Maximum Likelihood Estimates

                   Parameter   Standard StdErr                       Hazard
    Parameter DF   Estimate      Error  Ratio Chi-Square Pr > ChiSq  Ratio

    prestige   1   -0.07865    0.00615  0.972  163.4016    <.0001    0.924
    logsal     1   -0.59694    0.14249  1.245   17.5509    <.0001    0.550
    ed         1    0.18663    0.04604  1.595   16.4301    <.0001    1.205
    seq        1    0.26911    0.03057  1.038   77.5025    <.0001    1.309
```

Although we have corrected the standard errors for dependence, we have done nothing to improve the coefficients. Not only are they subject to attenuation bias (which we shall correct in the next session), they are also less than fully efficient. That means that the *true* standard errors are larger than they need to be. This inefficiency can be corrected using methods similar to those of Wei and colleagues (1989). The basic idea is to get coefficient estimates for each successive event (just as we did in Output 8.1) and then calculate a weighted average of those coefficients using the robust covariance matrix to calculate optimal weights.

The trick is to do this all in one run of PROC PHREG. (A SAS macro called WLW, which automates the following steps, can be downloaded at www.pauldallison.com.) The first step is to create interactions between each covariate and a set of dummy variables that indicate whether it's the first event, the second event, and so on. This can most easily be done in a

DATA step by creating arrays to hold the new variables and then defining the interactions in a DO loop:

```
DATA job;
  SET my.jobmult;
ARRAY p (*) p1-p10;
ARRAY s (*) s1-s10;
ARRAY e (*) e1-e10;
DO i=1 TO 10;
  p(i)=prestige*(seq=i);
  s(i)=logsal*(seq=i);
  e(i)=ed*(seq=i);
END;
RUN;
```

We then fit a model with PROC PHREG using all 30 of these covariates (plus SEQ itself), while adjusting the standard errors with the COVSANDWICH option. Three TEST statements are then used to calculate a weighted average for each set of coefficients:

```
PROC PHREG DATA=job COVS(AGGREGATE);
   MODEL duration*event(0)=p1-p10 s1-s10 e1-e10 seq;
   ID id;
   Prestige: TEST p1,p2,p3,p4,p5,p6,p7,p8,p9,p10 / AVERAGE;
   Salary: TEST s1,s2,s3,s4,s5,s6,s7,s8,s9,s10 / AVERAGE;
   Education: TEST e1,e2,e3,e4,e5,e6,e7,e8,e9,e10 / AVERAGE;
RUN;
```

There's little point in looking at the 30 coefficients in the fitted model. What's important are the weighted averages reported in Output 8.6. These estimates are similar in magnitude to those in Output 8.5 (although the weighted coefficient for salary is 43 percent larger than the unweighted one). The standard errors are slightly lower, which is what you hope for with a method that is theoretically more efficient.

Output 8.6 *PROC PHREG Results with Robust Standard Errors*

```
                    Average Effect for Test Prestige

                      Standard
        Estimate       Error        z-Score      Pr > |z|

        -0.0730        0.0057       -12.7635      <.0001


                    Average Effect for Test Salary

                      Standard
        Estimate       Error        z-Score      Pr > |z|

        -0.8601        0.1117       -7.6995       <.0001


                    Average Effect for Test Education

                      Standard
        Estimate       Error        z-Score      Pr > |z|

        0.2534         0.0332       7.6352        <.0001
```

Random-Effects Models

A second approach to the problem of dependence not only corrects the standard errors and test statistics, but it also corrects for some or all of the bias in the coefficients caused by unobserved heterogeneity. The basic idea is to formulate a model that explicitly introduces a disturbance term representing unobserved heterogeneity. Models of this sort are sometimes described as *subject-specific*.

Letting $h_{ij}(t)$ be the hazard for the *j*th event for individual *i* at time *t*, we can write

$$\log h_{ij}(t) = \alpha(t) + \boldsymbol{\beta}\mathbf{x}_{ij} + \varepsilon_i \tag{8.1}$$

where ε_i represents unobserved heterogeneity. Notice that ε is subscripted by *i* but not by *j*, indicating that the unobserved component is constant from one job to the next. At this point, there are two ways to proceed. In the next section, we will consider a fixed-effects approach. Here we shall assume that ε_i is a random variable with a specified distribution, independent of $\mathbf{x}_{ij}$. This leads to random-effects or frailty models that can be estimated by maximum likelihood or partial likelihood (Klein, 1992;

McGilchrist, 1993). When events are repeated, such models are well identified and are not highly sensitive to the choice of a distribution for ε.

Unfortunately, SAS has no procedure that is explicitly designed to estimate models like this. However, with a bit of programming, it is possible to estimate parametric random-effects models using the NLMIXED procedure. PROC NLMIXED does maximum likelihood estimation of non-linear mixed models (that is, models that incorporate both random effects and fixed effects). For example, we can estimate a Weibull random-effects model, which sets $\alpha(t) = \mu + \alpha \log t$ in equation (8.1). To estimate this model in PROC NLMIXED, you must specify the log-likelihood for a single individual. Let t_i be the event or censoring time for individual i, and let $\delta_i = 1$ if individual i is uncensored and 0 if i is censored. Define $\lambda_i = \exp(\boldsymbol{\beta}\mathbf{x}_i + \varepsilon_i)$, where ε_i has a normal distribution with a mean of 0 and a variance σ^2. It can be shown (using formulas in Chapter 4) that the contribution to the log-likelihood (conditional on ε) for a single individual is

$$\log L_i = -\lambda_i t_i^{(\alpha+1)} + \delta_i(\log(\alpha+1) + \alpha\log t_i + \log\lambda_i).$$

Here is the corresponding PROC NLMIXED program:

```
PROC NLMIXED DATA=jobmult;
 lambda=exp(b0+bed*ed+bpres*prestige+bsal*logsal+bseq*seq+e);
 ll=-lambda*duration**(alpha+1)+ event*(LOG(alpha+1)+
    alpha*LOG(duration)+LOG(lambda));
 MODEL duration~GENERAL(ll);
 RANDOM e~NORMAL(0,s2) SUBJECT=id;
 PARMS b0=1 bed=0 bpres=0 bsal=0 btime=0 s2=1 alpha=0;
RUN;
```

Because PROC NLMIXED can estimate such a wide range of non-linear models, it is necessary to specify the model in considerably more detail than in most other procedures. The statement beginning with LAMBDA defines $\lambda_i = \exp(\boldsymbol{\beta}\mathbf{x}_i + \varepsilon_i)$. In PROC NLMIXED, every parameter must be assigned a name. It's advisable to give the coefficients names that contain the corresponding variable names so that the output can be more easily interpreted. The E at the end of the statement corresponds to ε, the random effect. The statement beginning with LL defines the log-likelihood. Note that EVENT is the censoring indicator, equivalent to δ_i.

The MODEL statement sets DURATION as the response variable, with a distribution corresponding to the log-likelihood LL. The RANDOM statement declares E to be a normally distributed random variable with a mean of 0 and a variance S2 (which is just the name of the parameter to be estimated). The SUBJECT option says that E is a different random variable for each value of the ID variable. The PARMS statement sets the starting values for the parameters. If you omit the PARMS statement, the default starting values are set to 1 for all parameters, which often doesn't work. As shown here, I usually get results by setting all coefficients to 0, setting the intercept to 1 and the variance parameter to 1. But if you run into problems, you may do better by using coefficient estimates from PROC PHREG as starting values.

The key results are shown in Output 8.7. The coefficient for PRESTIGE (BPRES) is about the same as in Output 8.5, with similar standard error. LOGSAL has a coefficient (BSAL) that is substantially larger in magnitude than in Output 8.5 and with a slightly smaller standard error. The big change is in the effect of education (ED). The coefficient in Output 8.5 was positive and highly significant, but here the coefficient (BED) is negative and almost significant at the .05 level. Last, the coefficient for SEQ, which was large and highly significant in Output 8.5, is now much smaller and not quite statistically significant. All of these discrepancies between Outputs 8.5 and 8.7 are due to the presence of the random error term ε. If you remove E and the RANDOM statement from the PROC NLMIXED program, you get estimates for a conventional Weibull model that are very similar to those in Output 8.5.

Besides the coefficient estimates, we also get an estimate of S2, the variance of ε, which is large and highly significant. This tells us that there is definitely unobserved heterogeneity across persons or, equivalently, that there is dependence among the repeated observations. The last line is an estimate of α, the coefficient of log t. This can be interpreted by saying that each 1 percent increase in time is associated with a 2.2 percent increase in the hazard of job termination.

Output 8.7   *PROC NLMIXED Results for Weibull Model with Random Effects*

Parameter	Estimate	Standard Error	DF	t Value	Pr > |t|
b0	2.5528	1.3258	99	1.93	0.0570
bpres	-0.08032	0.008202	99	-9.79	<.0001
bsal	-0.9810	0.1523	99	-6.44	<.0001
bed	-0.2265	0.1178	99	-1.92	0.0575
bseq	0.06754	0.03637	99	1.86	0.0663
s2	6.7986	1.5026	99	4.52	<.0001
alpha	2.1943	0.1627	99	13.48	<.0001

Fixed-Effects Models

Instead of treating ε_i in equation (8.1) as a set of random variables with a specified probability distribution, we can treat it as representing a set of fixed constants (that is, a fixed-effects model). The attraction of this approach is that, rather than assuming that ε_i is independent of $\mathbf{x}_i$, we allow ε_i to be freely correlated with $\mathbf{x}_i$. That implies that the estimates actually control for all stable, unobserved characteristics of the individual.

How can we estimate such a model? One possibility is to put dummy variables in the model for all individuals (except one). This method works well for linear regression models but not for Cox regression, which is subject to something called *incidental parameters* bias (Allison, 2002). When the average number of intervals per person is less than three, regression coefficients are inflated by approximately 30 to 90 percent, depending on the level of censoring (a higher proportion of censored cases produces greater inflation).

Fortunately, there is a simple alternative method that does the job very well. First, we modify equation (8.1) to read

$$\log h_{ij}(t) = \alpha_i(t) + \boldsymbol{\beta}\mathbf{x}_{ij}.$$

In this equation, the fixed effect ε_i has been absorbed into the unspecified function of time, which is now allowed to vary from one individual to another. Thus, each individual has her own hazard function, which is considerably less restrictive than allowing each individual to have her own constant.

This model can be estimated by partial likelihood using the method of stratification, discussed in Chapter 5. Stratification allows different subgroups to have different baseline hazard functions, while constraining the coefficients to be the same across subgroups. We can implement this method in PROC PHREG by using the STRATA statement, with an identification variable that distinguishes individuals. This is called the *fixed-effects partial likelihood* (FEPL) method. For the job duration data set, the SAS statements are

```
PROC PHREG DATA=jobmult NOSUMMARY;
   MODEL duration*event(0)=prestige logsal seq;
   STRATA id;
RUN;
```

The NOSUMMARY option suppresses the information that is usually reported for each stratum. Otherwise, you get a line of output for every individual in the sample.

Notice that the variable ED is not included as a covariate. One drawback of the FEPL method is that it can only estimate coefficients for those covariates that vary across (or within) the successive spells for each individual. And because education does not vary over time for this sample, its effect cannot be estimated. On the other hand, the fixed-effect method implicitly controls not only for education *but for all constant covariates,* regardless of whether they are measurable and regardless of whether they are correlated with the measured covariates. In other words, this method controls for things like race, ethnicity, sex, religion, region of origin, personality—anything that's stable over time.

As Output 8.8 shows, a fixed-effects analysis for the job duration data yields results that are similar to those for the random-effects model in Output 8.6, except that the effect of SEQ is even smaller.

Output 8.8 *Results of Fixed-Effects Partial Likelihood for Repeated Job Durations*

Parameter	DF	Parameter Estimate	Standard Error	Chi-Square	Pr > ChiSq	Hazard Ratio
prestige	1	-0.05594	0.00979	32.6427	<.0001	0.946
logsal	1	-0.86566	0.17432	24.6616	<.0001	0.421
seq	1	0.02073	0.03941	0.2769	0.5988	1.021

Fixed-effects partial likelihood was proposed by Chamberlain (1985), who expressed reservations about its use when the number of intervals varied across individuals and when the censoring time (the time between the start of an interval and the termination of observation) depended on the lengths of preceding intervals. Both of these conditions exist for the job duration data set and for most other applications to repeated events. Whatever the theoretical merit of these concerns, however, my own (1996) Monte Carlo simulations have convinced me that there is little or no problem in practice.

There is one exception to that conclusion: the FEPL method does not do a good job of estimating the effect of the number of previous events, such as the SEQ variable in this example. Specifically, FEPL tends to yield negative estimates for the number of prior events, even when there is no real effect. On the other hand, conventional partial likelihood produces estimates for the number of prior events that are much more strongly biased but in the opposite direction (as we found in Output 8.3).

As already noted, a major advantage of the FEPL method is that the unobserved disturbance term is allowed to be correlated with the measured covariates. However, when the disturbance term is *not* correlated with any of the covariates, the random-effects approach produces more efficient estimates (that is, with smaller standard errors). This differential efficiency is especially great when the average number of events per individual is less than two. The reason is that the FEPL method excludes two types of individuals from the partial likelihood function:

- those with no events (that is, those with only a single censored spell)
- those with one uncensored spell and one censored spell, *if* the censored spell is shorter than the uncensored spell.

One circumstance in which the unobservable characteristics of individuals are not correlated with their measured characteristics is a randomized experiment. But with nonexperimental data, it's much safer to assume that they *are* correlated.

In sum, the FEPL method is the preferred technique for repeated events whenever

- the data do not come from a randomized experiment
- interest is centered on covariates that vary across intervals for each individual
- most individuals have at least two events
- there is a reasonable presumption that the process generating events is invariant over time.

If the data are produced by a randomized experiment or if the main interest is in covariates that are constant for each individual, a random-effects method is preferable.

Specifying a Common Origin for All Events

Another issue that arises in the analysis of repeated events is whether the hazard varies as a function of time since the last event or as a function of time since the process began. All the analyses so far have made the former assumption, that the clock is reset to 0 whenever an event occurs. But in many applications, it is reasonable to argue that the hazard depends on the time since the individual first became at risk, regardless of how many intervening events have occurred. For example, the hazard for a job change may depend on the time in the labor force rather than the time in the current job.

PROC PHREG makes it easy to specify models in which the hazard depends on a single origin for all of an individual's events. As before, we create a separate record for each interval for each individual. But instead of a single variable containing the length of the interval, the record must contain a starting time (measured from the common origin) and a stopping time (also measured from the origin). Output 8.9 shows what the records

look like for the job duration data set (omitting the covariates). Here is the
program that produced this data set and output:

```
DATA strtstop;
  SET jobmult;
  RETAIN stop;
  IF seq=1 THEN stop=0;
  start=stop;
  stop=duration+stop;
PROC PRINT;
  VAR id event seq start stop;
RUN;
```

Output 8.9 *First 20 Cases for Job Duration Data with Start and Stop Times*

OBS	ID	EVENT	SEQ	START	STOP
1	1	1	1	0.0000	2.3575
2	1	1	2	2.3575	6.9029
3	1	1	3	6.9029	7.9893
4	1	1	4	7.9893	11.5786
5	1	1	5	11.5786	16.2462
6	1	0	6	16.2462	20.0000
7	2	0	1	0.0000	20.0000
8	3	1	1	0.0000	7.3753
9	3	1	2	7.3753	18.6385
10	3	0	3	18.6385	20.0000
11	4	1	1	0.0000	1.1436
12	4	1	2	1.1436	1.9315
13	4	1	3	1.9315	2.9752
14	4	1	4	2.9752	3.2098
15	4	1	5	3.2098	4.0081
16	4	1	6	4.0081	5.2249
17	4	1	7	5.2249	6.1409
18	4	1	8	6.1409	6.4583
19	4	1	9	6.4583	7.1156
20	4	1	10	7.1156	8.8192

The model is then specified by the following program:

```
PROC PHREG DATA=strtstop COVSANDWICH(AGGREGATE);
  MODEL (start,stop)*event(0)=prestige logsal ed seq;
  ID id;
RUN;
```

Notice that I have also used the COVSANDWICH option to adjust for
possible dependence among the multiple events for each person. The
results in Output 8.10 are quite similar to those in Output 8.5, which reset
the origin for the hazard function at each job termination, although the

coefficient magnitudes are somewhat smaller. These are essentially estimates of the counting process model proposed by Andersen and Gill (1982), although they did not correct the standard errors for dependence. If we stratified by SEQ rather than included it as a covariate, we would get estimates of the total time model described by Prentice and colleagues (1981).

Output 8.10 *PROC PHREG Estimates with a Common Origin for Pooled Job Durations*

Parameter	DF	Parameter Estimate	Standard Error	StdErr Ratio	Chi-Square	Pr > ChiSq	Hazard Ratio
prestige	1	-0.05223	0.00382	0.671	187.0645	<.0001	0.949
logsal	1	-0.37332	0.08465	0.710	19.4493	<.0001	0.688
ed	1	0.12815	0.03439	1.170	13.8827	0.0002	1.137
seq	1	0.17951	0.02104	0.709	72.7975	<.0001	1.197

An alternative approach is to use the stop times for each observational record but not the start times. If this is done, however, it is essential that one also stratify on the sequence number, as in the marginal model of Wei and colleagues (1989). Otherwise, the same individual could appear more than once in the same risk set. For example, we can use

```
PROC PHREG DATA=strtstop COVSANDWICH(AGGREGATE);
  MODEL stop*event(0)=prestige logsal ed;
ID id;
STRATA seq;
RUN;
```

This approach produces the results in Output 8.11, which are quite similar to those in Output 8.10.

Output 8.11 *PROC PHREG with Common Origin and No Adjustment of Start Times*

Parameter	DF	Parameter Estimate	Standard Error	StdErr Ratio	Chi-Square	Pr > ChiSq	Hazard Ratio
prestige	1	-0.06167	0.00628	0.987	96.5662	<.0001	0.940
logsal	1	-0.52963	0.13594	1.123	15.1788	<.0001	0.589
ed	1	0.13875	0.04859	1.567	8.1543	0.0043	1.149

One disadvantage of specifying a single time origin for all events is that you cannot then do fixed-effects partial likelihood. That is because each individual is treated as a separate stratum in FEPL. For each of those strata, a common time origin implies that there is only one observation in the risk set at any given point in time, so the partial likelihood becomes degenerate. Furthermore, although it's theoretically possible to fit a random-effects model with a common time origin, I am not aware of any software that will do this.

Repeated Events for Discrete-Time Maximum Likelihood

In Chapter 7, we saw how to use PROC LOGISTIC to estimate models for discrete-time survival data using the method of maximum likelihood. These methods can easily be extended to handle repeated events. In fact, there are even more methods available for discrete-time maximum likelihood, and some of those methods are simpler to implement than the ones that we have already discussed in this chapter.

Before discussing the analytic methods, we must first consider the construction of the data set. In Chapter 7, for non-repeated events, the data set consisted of one record for each unit of time that each individual was observed, up to and including the time unit in which an event occurred. But once an event occurred, no more observational records were created. When events are repeatable, however, you don't stop with the first event occurrence. An observational record is created for each time unit that the individual is observed, regardless of how many events occur before or after.

Once the data set is created, you can simply do a binary regression using PROC LOGISTIC, either with the default logit link (for truly discrete time) or the complementary log-log link (for grouped continuous-time data). Unlike in Chapter 7, however, methods are needed to correct for dependence among the repeated observations. These methods are discussed in detail in Chapter 3 of my book *Fixed Effects Regression Methods for Longitudinal Data Using SAS* (SAS Institute, 2005).

Here is an example of how to do it for the job duration data. In the following DATA step, I first convert the DURATION variable into discrete years by using the CEIL function (which rounds up to the next higher integer). Then, using a DO loop, I produce a set of multiple records for each person, for a total of 2,072 person years:

```
DATA discrete;
  SET jobmult;
  durdis=CEIL(duration);
  DO time=1 TO durdis;
    term=0;
    IF time=durdis THEN term=event;
```

```
        OUTPUT;
      END;
   RUN;
```

To implement the method of robust standard errors, you can use PROC SURVEY LOGISTIC with the CLUSTER statement:

```
PROC SURVEYLOGISTIC DATA=discrete;
   MODEL term(DESC)=prestige logsal ed seq time time*time;
   CLUSTER id;
RUN;
```

Notice that, unlike the PROC PHREG models, it's important to include the time since the last event as a covariate. (Actually, I've allowed for a quadratic function by including both TIME and TIME squared.) In PROC PHREG, dependence on time is automatically allowed via the $\alpha(t)$ function. As expected, the results in Output 8.11 are very similar to those in Output 8.5. If you prefer to estimate a proportional hazards model rather than a logit model, just put the / LINK=CLOGLOG option in the MODEL statement.

Output 8.11 *PROC SURVEYLOGISTIC with Robust Standard Errors*

Parameter	DF	Estimate	Standard Error	Wald Chi-Square	Pr > ChiSq
Intercept	1	-0.4169	0.6629	0.3954	0.5295
prestige	1	-0.0894	0.00840	113.2848	<.0001
logsal	1	-0.6253	0.1807	11.9773	0.0005
ed	1	0.2114	0.0602	12.3411	0.0004
seq	1	0.3317	0.0385	74.2487	<.0001
time	1	0.4012	0.0857	21.9219	<.0001
time*time	1	-0.0162	0.00492	10.8633	0.0010

Random-effects models (for either link function) can be estimated with PROC NLMIXED, although the syntax is a little simpler than what we used for the Weibull model earlier in this chapter. Even easier is PROC GLIMMIX, which has a much simpler syntax than PROC NLMIXED. Here is the code:

```
PROC GLIMMIX DATA=discrete METHOD=QUAD;
   MODEL term=prestige logsal ed seq time
     /DIST=BIN SOLUTION LINK=LOGIT;
     RANDOM INTERCEPT / SUBJECT=id;
RUN;
```

By default, PROC GLIMMIX does pseudo-likelihood estimation, which is very fast but may not be sufficiently accurate for binary outcomes. However, you can force it to do true maximum likelihood estimation by specifying METHOD=QUAD in the MODEL statement. This option uses Gaussian quadrature to calculate the likelihood function, the same method used by PROC NLMIXED.

Results in Output 8.12 are very similar to those produced by PROC NLMIXED in Output 8.7. The covariance parameter of 9.2179 is the estimated variance of ε_i.

Output 8.12 *PROC GLIMMIX Estimates for the Random-Effects Model*

```
                    Covariance Parameter Estimates


                                            Standard
            Cov Parm     Subject   Estimate    Error

            Intercept    id          9.2179   2.6330



                    Solutions for Fixed Effects

                          Standard
  Effect       Estimate     Error     DF    t Value    Pr > |t|

  Intercept      3.0191    1.6496      98      1.83      0.0703
  prestige      -0.09723   0.01266   1967     -7.68      <.0001
  logsal        -0.9589    0.2514    1967     -3.81      0.0001
  ed            -0.2857    0.1532    1967     -1.86      0.0624
  seq            0.08710   0.05818   1967      1.50      0.1345
  time           1.2281    0.1275    1967      9.63      <.0001
  time*time     -0.04313   0.005883  1967     -7.33      <.0001
```

To estimate a fixed-effects model, simply use PROC LOGISTIC with the STRATA statement, as follows:

```
PROC LOGISTIC DATA=discrete;
  MODEL term(DESC)=prestige logsal ed seq time time*time;
  STRATA id;
RUN;
```

However, unlike the robust standard error method, the fixed-effects method is only available for the logit link, not the complementary log-log link. Results in Output 8.13 are very similar to those in Output 8.8. Note that the coefficient for education cannot be estimated in a fixed-effects model because that variable does not vary over time.

Output 8.13 *Fixed-Effects Estimates with PROC LOGISTIC*

```
               Analysis of Maximum Likelihood Estimates

                              Standard        Wald
  Parameter    DF   Estimate     Error   Chi-Square   Pr > ChiSq

  prestige      1    -0.0762    0.0131     33.5857       <.0001
  logsal        1    -1.0499    0.2648     15.7198       <.0001
  ed            0   -4.05E-6         .          .            .
  seq           1   -0.00772    0.0571      0.0183       0.8924
  time          1     1.3716    0.1400     96.0533       <.0001
  time*time     1    -0.0346   0.00765     20.4496       <.0001
```

Besides these methods, one can also adjust for dependence using the method of generalized estimating equations (GEE), which is available in PROC GENMOD. Invoked with the REPEATED statement, this method produces robust standard error estimates using the same method that is used by PROC PHREG. However, GENMOD also produces GEE estimates of the coefficients, which are more statistically efficient than conventional logistic regression coefficients. Although GEE estimates have less sampling variability than conventional estimates, they do not adjust for bias due to unobserved heterogeneity.

Specifying the hazard as a function of time since a common origin is easy with any of these models for discrete-time data. That's because time is treated like any other variable on the right-hand side of the equation. So you simply create another variable, which is the time since origin, and include that in the model. You can also include the time since the last event in the same model. And you can specify nonlinear functions of either of these variables.

GENERALIZED R^2

People who do a lot of linear regression tend to become attached to R^2 as a measure of how good their models are. When they switch to PROC LIFEREG or PROC PHREG, they may experience withdrawal symptoms because no similar statistic is reported. In this section, I show how a generalized R^2 can be easily calculated from statistics that *are* reported. Before doing that, I want to caution readers that R^2 is not all it's cracked up to be, regardless of whether it's calculated for a linear model or a proportional hazards model. In particular, R^2 does *not* tell you anything about how appropriate the model is for the data. You can obtain an R^2 of only .05 for a model whose assumptions are perfectly satisfied by the data and whose coefficients are precisely unbiased. Similarly, an R^2 of .95 does not protect you against severe violations of assumptions and grossly biased coefficients. All R^2 tells you is how well you can predict the dependent variable with the set of covariates. And even for prediction, some authorities argue that the standard error of the estimate is a more meaningful and useful measure.

Still, other things being equal, a high R^2 is definitely better than a low R^2, and I happen to be one of those who miss it if it's not there. With that in mind, let's see how we can calculate one for survival models. Unfortunately, there's no consensus on the best way to calculate an R^2 for nonlinear models. In my opinion, many of the R^2s that are reported by some widely used packages are virtually worthless. The statistic that I describe here was proposed by Cox and Snell (1989) and was also one of three statistics endorsed by Magee (1990). It's the same statistic that is available as an option for PROC LOGISTIC.

Let G^2 be the likelihood ratio chi-square statistic for testing the null hypothesis that all covariates have coefficients of 0. This G^2 is reported directly by PROC PHREG and PROC LOGISTIC. For PROC LIFEREG, you must calculate it yourself by fitting models both with and without the covariates and then taking twice the positive difference in the log-likelihoods. With that statistic in hand, you can calculate the R^2 as

$$R^2 = 1 - \exp\left(\frac{-G^2}{n}\right) \tag{8.2}$$

where n is the sample size. If you are using a method that breaks the individual's time to event into multiple records (for example, the piecewise exponential model or the maximum likelihood logistic model), the n should be the number of individuals, not the number of observational records.

The justification for this formula is simple. For an ordinary linear regression model with a normal error term, it's possible to calculate a likelihood ratio chi-square statistic for the null hypothesis that all coefficients are 0 (although what's usually reported is an F statistic). That chi-square statistic is related to the standard R^2 by the formula in equation (8.2). By analogy, we use the same formula for nonlinear models.

You can use this statistic for any regression model estimated by maximum likelihood or partial likelihood. Unlike the linear model, however, it cannot be interpreted as a proportion of variation in the dependent variable that is explained by the covariates. It's just a number between 0 and 1 that is larger when the covariates are more strongly associated with the dependent variable. Nevertheless, it seems to behave in similar ways to the usual R^2. In samples with no censored data, I have compared OLS linear regression models for the logarithm of time with various accelerated failure time models estimated by maximum likelihood and with Cox models estimated by partial likelihood. In all cases that I have examined, the generalized R^2s from the likelihood-based procedures are similar in magnitude to the R^2 from the ordinary regression model.

Here are some examples. In Chapter 4, "Estimating Parametric Regression Models with PROC LIFEREG," we used PROC LIFEREG to estimate a Weibull model for the 432 cases in the recidivism data set, and we calculated a likelihood ratio chi-square statistic of 33.48 for the test that all coefficients are 0. Applying the formula above, we get an R^2 of .0746. For the same data set, we used partial likelihood in Chapter 5 to estimate a proportional hazards model. As shown in Output 5.1, the likelihood ratio chi-square statistic is 33.13, so the R^2 is virtually identical. For the 65 heart transplant cases, the PROC PHREG model in Output 5.3 had a likelihood ratio chi-square statistic of 16.6, yielding an R^2 of .23.

SENSITIVITY ANALYSIS FOR INFORMATIVE CENSORING

In Chapters 2 and 6, we discussed the relatively intractable problem of informative censoring. Let's quickly review the problem. Suppose that just before some particular time t, there are 50 individuals who are still at risk of an event. Of those 50 individuals, 5 are censored at time t. Suppose further that 20 of the 50 at risk have covariate values that are identical to those of the 5 who are censored. We say that censoring is *informative* if the 5 who are censored are a biased subsample of the 20 individuals with the same covariate values. That is, they have hazards that are systematically higher or lower than those who were not censored. Informative censoring can lead to parameter estimates that are seriously biased.

When censoring is random (that is, not under the control of the investigator), it's usually not difficult to imagine scenarios that would lead to informative censoring. Suppose, for example, that you're studying how long it takes rookie policemen to be promoted, and those who quit are treated as censored. It doesn't take much insight to suspect that those who quit before promotion have, on average, poorer prospects for promotion than those who stay. Unfortunately, there's no way to test this hypothesis. You can compare the performance records and personal characteristics of those who quit and those who stayed, but these are all things that would probably be included as covariates in your model. Remember that what we're concerned about is *residual* informativeness, after the effects of covariates have been taken into account. And even if we could discriminate between informative and noninformative censoring, there are no standard methods for handling informative censoring. The best that can be done by way of correction is to include as covariates any factors that are believed to affect both event times and censoring times.

There is, however, a kind of sensitivity analysis that can give you some idea of the possible impact that informative censoring might have on your results. The basic idea is to redo the analysis under two extreme assumptions about censored cases. One assumption is that censored observations experience events immediately after they are censored. This corresponds to the hypothesis that censored cases are those that tend to be at high risk of an event. The opposite assumption is that censored cases have longer times to events than anyone else in the sample. Obviously, this corresponds to the hypothesis that censored cases are those that tend to be at low risk of an event.

This is the general strategy. Some care needs to be taken in implementation, however. Let's consider the LEADERS data set that we analyzed extensively in Chapter 6. There were four outcomes, as coded in the variable LOST:

Code	Frequency	Reason
0	115	leader still in power at the end of the study
1	165	leader left office by constitutional means
2	27	leader died of natural causes
3	165	leader left office by nonconstitutional means

We saw that types 1 and 3 were similar in many respects. Output 6.12 displayed estimates for a Cox model that combined types 1 and 3 but that treated types 0 and 2 as censoring. Now we want to see how sensitive those estimates in Output 6.12 are to possible informative censoring.

The censoring that occurred because the study ended is, in fact, random censoring because the leaders began their spells in power at varying points in time. It's possible that those who entered power later had higher or lower risks of exit than those who entered earlier. Nonetheless, we can control for this kind of random censoring by including the START time as a covariate, which we did in Output 6.12. We can't do anything like that for deaths, however, so our sensitivity analysis will focus on the 27 cases that have a value of 2 for the LOST variable.

The first reanalysis assumes that if the 27 people had not died in a particular year, they would have *immediately* been removed from power by constitutional or nonconstitutional means. We can accomplish this reanalysis by removing the value 2 from the list of censored values in the MODEL statement, thereby treating it as if it were a 1 or a 3:

```
PROC PHREG DATA=leaders;
   MODEL years*lost(0)=manner age start military conflict
         loginc literacy / TIES=EXACT;
   STRATA region;
RUN;
```

The second reanalysis assumes that if those 27 leaders had not died of natural causes, they would have remained in power at least as long as anyone else in the sample. We still treat them as censored, but we change their censoring time to the largest event time in the sample, which, in this case, is 24 years. Because we're doing a partial likelihood analysis, we could change it to any number greater than or equal to 24, and the result would be the same. If we were using PROC LIFEREG, which uses the exact times of all observations, we might want to change the censoring time to the longest observed time, either censored or uncensored. The longest censored time in this sample is 27 years.

Here's the SAS code for changing the censoring times to 24 and then running the model:

```
DATA leaders2;
   SET leaders;
   IF lost=2 THEN years=24;
PROC PHREG DATA=leaders2;
   MODEL years*lost(0,2)=manner age start military conflict
         loginc literacy / TIES=EXACT;
   STRATA region;
RUN;
```

Compare the results for these two models, shown in Output 8.14, with those in Output 6.12. For each covariate, the two new estimates bracket the original estimate. The biggest difference is in the effect of age, which gets larger in the top panel but much smaller in the bottom panel, to the point where it is no longer significant at the .05 level. This should not be surprising because AGE was the only variable of any importance in the model for deaths due to natural causes (Output 6.7). For the most part, however, the results in Output 8.14 are reassuring. We can be reasonably confident that treating natural deaths as noninformative censoring has no appreciable affect on the conclusions.

Output 8.14 *Sensitivity Analysis of LEADERS Data Set for Informative Censoring*

Censored Cases Treated as Uncensored Cases

Parameter	DF	Parameter Estimate	Standard Error	Chi-Square	Pr > ChiSq	Hazard Ratio
manner	1	0.38528	0.15490	6.1868	0.0129	1.470
age	1	0.02307	0.00547	17.8215	<.0001	1.023
start	1	-0.01774	0.00816	4.7251	0.0297	0.982
military	1	-0.22637	0.16041	1.9913	0.1582	0.797
conflict	1	0.12992	0.13073	0.9876	0.3203	1.139
loginc	1	-0.18263	0.08235	4.9181	0.0266	0.833
literacy	1	0.0006600	0.00320	0.0426	0.8366	1.001

Censoring Times Recoded as 24

Parameter	DF	Parameter Estimate	Standard Error	Chi-Square	Pr > ChiSq	Hazard Ratio
manner	1	0.31912	0.15669	4.1479	0.0417	1.376
age	1	0.01001	0.00568	3.1128	0.0777	1.010
start	1	-0.01316	0.00826	2.5394	0.1110	0.987
military	1	-0.14295	0.16305	0.7686	0.3806	0.867
conflict	1	0.20979	0.13708	2.3421	0.1259	1.233
loginc	1	-0.27022	0.08939	9.1382	0.0025	0.763
literacy	1	0.00378	0.00334	1.2819	0.2575	1.004

In interpreting output like this, you should remember that these are worst-case scenarios, and it's unlikely that either of the extremes is an accurate depiction of reality. The usual estimates are still your best guess of what's really going on. It's also worth noting that for many applications, one of these extremes may be much more plausible than the other. Naturally, you'll want to focus your attention on the extreme that is most sensible.

CHAPTER **9**
A Guide for the Perplexed

HOW TO CHOOSE A METHOD

In this book, we've examined many different approaches to the analysis of survival data: Kaplan-Meier estimation, log-rank tests, accelerated failure time models, piecewise exponential models, Cox regression models, logit models, and complementary log-log models. Each of these methods is worth using in some situations. Along the way I have tried to point out their relative advantages and disadvantages, but those discussions are scattered throughout the book. It's not easy to keep all the points in mind when designing a study or planning an analysis.

Many readers of early versions of this book urged me to provide a concluding roadmap for choosing a method of survival analysis. Although I give this kind of advice all the time, I do so here with some hesitation. While statisticians may have great consensus about the characteristics of various statistical methods, the choice among competing methods is often very personal, especially when dealing with methods as similar in spirit and results as those presented here. Five equally knowledgeable consultants could easily give you five different recommendations. I'm going to present some rules of thumb that I rely on myself in giving advice, but please don't take them as authoritative pronouncements. Use them only to the degree that you find their rationales persuasive.

Make Cox Regression Your Default Method

Given the relative length of Chapter 5, "Estimating Cox Regression Models with PROC PHREG," it will come as no surprise that I have a strong preference for Cox regression using PROC PHREG. This particular method

- is more robust than the accelerated failure time methods
- has excellent capabilities for time-dependent covariates
- handles both continuous-time and discrete-time data
- allows for late entry and temporary exit from the risk set

■ has a facility for nonparametric adjustment of nuisance variables (stratification).

PROC PHREG can also do log-rank tests (using a single dichotomous covariate). PROC PHREG will even do Kaplan-Meier estimation (by fitting a model with no covariates and using the BASELINE statement to produce a table of survival probabilities).

Beyond these intrinsic advantages, Cox regression has the considerable attraction of being widely used, accepted, and understood. What's the point of choosing a marginally superior method if your audience is confused or skeptical and you have to waste valuable time and space with explanation and justification? For better or worse (mostly for better), Cox regression has become the standard, and it makes little sense to choose a different method unless you have a good reason for doing so. Having said that, I'll now suggest some possible good reasons. They are presented in the form of questions that you should ask yourself when deciding on a method.

Is the Sample Large with Heavily Tied Event Times?

As we saw in Chapter 5, PROC PHREG *can* deal with situations in which there are many events occurring at the same recorded times using either the DISCRETE method or the EXACT method. Unfortunately, those options can take a great deal of computer time that increases rapidly with sample size. If you have a sample of 10,000 observations with only 10 distinct event times, you can expect to wait a long time before you see any output. Why wait when the alternatives are so attractive? The complementary log-log and logit methods of Chapter 7 estimate *exactly* the same models as the EXACT and DISCRETE methods with statistical efficiency that is at least as good as that provided by PROC PHREG.

Do You Want to Study the Shape of the Hazard Function?

In some studies, one of the major aims is to investigate hypotheses about the dependence of the hazard on time. Cox regression is far from ideal in such situations because it treats the dependence on time as a nuisance function that cancels out of the estimating equations. You can still produce graphs of the baseline survival and hazard functions, but those graphs don't provide direct tests of hypotheses.

With PROC LIFEREG, on the other hand, you can produce formal hypothesis tests that answer the following sorts of questions:

■ Is the hazard constant over time?

■ If not constant, is the hazard increasing, decreasing, or non-monotonic?

■ If increasing (or decreasing), is the rate of change going up or down?

All of these questions are addressed within the context of smooth parametric functions. If that seems too restrictive, you can get much more flexibility with the piecewise exponential, or with maximum likelihood estimation of the logit or complementary log-log models. With these models, the time scale is chopped into intervals, and the least restrictive models have a set of indicator variables to represent those intervals. Restrictions can easily be imposed to represent functions of almost any desired shape. A further advantage of these multiple-record methods is that two or more time axes can be readily introduced into a single model. A model for promotions, for example, could include the time since the last promotion, the time since initial employment by the firm, the time in the labor force, and the time since completion of education.

So if you want to study the dependence of the hazard on time, there are good alternatives to Cox regression. But whatever method you use, I urge you to be cautious in interpreting the results. As we saw in Chapter 8, "Heterogeneity, Repeated Events, and Other Topics," the hazard function is strongly confounded with uncontrolled heterogeneity, which makes hazards look like they are declining with time even when they are constant or increasing. As a result, any declines in the hazard function may be purely artifactual, a possibility that you can never completely rule out.

Do You Want to Generate Predicted Event Times or Survival Probabilities?

As we saw in Chapter 5, you can use output from the BASELINE statement in PROC PHREG to get predicted median survival times or survival probabilities for any specified set of covariates. Because the BASELINE statement produces a complete set of survivor function estimates, however, getting predicted values or 5-year survival probabilities for a large number of observations can be rather cumbersome. Furthermore, predicted median survival times may be unavailable for many or all of the observations if a substantial fraction of the sample is censored. With PROC LIFEREG, on the other hand, you can easily generate predicted median survival times (or any other percentile) for all observations using the OUTPUT statement. Using my PREDICT macro, you can also produce estimated survival probabilities for a specified survival time.

Do You Have Left-Censored Data?

The only SAS procedure that allows for left censoring is PROC LIFEREG. In principle, it's possible to adapt Cox regression to handle left censoring, but I know of no commercial program that does this.

CONCLUSION

While it's good to consider these questions carefully, you shouldn't become obsessed with making the *right* choice. For most applications, a reasonable case can be made for two or more of these methods. Furthermore, the methods are so similar in their underlying philosophy that they *usually* give similar results. When they differ, it's typically in situations where the evidence is not strong for any conclusion. If you have the time, it's always worthwhile trying out two or more methods on the same data. If they lead you to the same conclusions, then your confidence is increased. If they are discrepant, your confidence is appropriately reduced. If they are widely discrepant, you should carefully investigate the reasons for the discrepancy. Perhaps you made an error in setting up one method or the other. Maybe there's a serious peculiarity in the data. Search for outliers and examine residuals and influence diagnostics. Discrepancies are often a powerful indication that there is something important to be learned.

APPENDIX **1**
Macro Programs

INTRODUCTION

In the main text, I introduced two macros for carrying out certain auxiliary analyses with PROC LIFEREG. They are listed here with instructions and background information. They are also available on the Web at www.pauldallison.com.

Both of the macro programs use keywords for the required and optional parameters. Default values (if any) are given after the equal sign in the parameter list. Thus, you only need to specify parameters that differ from the default value, and you can specify these parameters in any order in the macro call.

THE LIFEHAZ MACRO

The LIFEHAZ macro produces parametric plots of hazard functions based on models fitted by PROC LIFEREG. In the LIFEREG procedure, you must specify OUTEST=*name1* in the PROC LIFEREG statement. You must also use the OUTPUT statement with OUT=*name2* and XBETA=*name3*. By default, the hazard is plotted for the mean value of the XBETA option (the linear predictor). If you want a plot for a specific observation, you must specify the observation number (OBSNO) when you invoke the macro. The macro requires SAS/GRAPH software.

To use the macro, you must first read it into your SAS session. This can be done by simply copying the text of the macro into the editor window and then running it. Alternatively, you can use the %INCLUDE statement to specify the name and location of the file containing the macro (which must have a .sas extension). Once the macro has been read in to SAS, you can use it repeatedly during your SAS session. For an example using the LIFEHAZ macro, see the section **Generating Predictions and**

Hazard Functions in Chapter 4, "Estimating Parametric Regression Models with PROC LIFEREG."

Parameters

OUTEST=	specifies the name of the data set produced by the OUTEST= option in PROC LIFEREG.
OUT=_LAST_	specifies the name of the data set produced by the OUT= option in PROC LIFEREG.
XBETA=	specifies the name used with the XBETA= option in the OUTPUT statement.
OBSNO=	specifies the sequence number of the observation for which you want plots of the hazard function. (The default method is to use the mean of the linear predictor to generate the hazards.)

Program

```
%macro lifehaz(outest=,out=,obsno=0,xbeta=lp);
/******************************************************************
Version 2.0 (9-14-01)

This version of LIFEHAZ works for SAS Release 6.12 through
Release 9.2.

******************************************************************/
data;
  set &outest;
  call symput('time',_NAME_);
run;
proc means data=&out noprint;
  var &time &xbeta;
  output out=_c_ min(&time)=min max(&time)=max mean(&xbeta)=mean;
run;
data;
  set &outest;
  call symput('model',_dist_);
  s=_scale_;
  d=_shape1_;
  _y_=&obsno;
  set _c_ (keep=min max mean);
  if _y_=0 then m=mean;
  else do;
    set &out (keep=&xbeta) point=_y_;
    m=&xbeta;
  end;
  inc=(max-min)/300;
  g=1/s;
```

```
  alph=exp(-m*g);
  _dist_=upcase(_dist_);
if _dist_='LOGNORMAL' or _dist_='LNORMAL'  then do;
  do t=min to max by inc;
  z=(log(t)-m)/s;
  f=exp(-z*z/2)/(t*s*sqrt(2*3.14159));
  Surv=1-probnorm(z);
  h=f/Surv;
  output;
  end;
end;
else if _dist_='GAMMA' then do;
  k=1/(d*d);
  do t=min to max by inc;
  u=(t*exp(-m))**(1/s);
  f=abs(d)*(k*u**d)**k*exp(-k*u**d)/(s*gamma(k)*t);
  Surv=1-probgam(k*u**d,k);
  if d lt 0 then Surv=1-Surv;
  h=f/Surv;
  output;
  end;
end;
else if _dist_='WEIBULL' or _dist_='EXPONENTIAL' or _dist_='EXPONENT'  then do;
  do t=min to max by inc;
  h=g*alph*t**(g-1);
  output;
  end;
end;
else if _dist_='LLOGISTIC' or _dist_='LLOGISTC' then do;
  do t=min to max by inc;
  h=g*alph*t**(g-1)/(1+alph*t**g);
  output;
  end;
end;
else put 'ERROR:DISTRIBUTION NOT FITTED BY LIFEREG';
run;
proc gplot;
  plot h*t / haxis=axis2 vaxis=axis1 vzero;
  symbol1 i=join v=none c=black;
  axis1 label=(f=titalic angle=90 'Hazard');
  axis2 label=(f=titalic justify=c 'time' f=titalic justify=c "&model");
run; quit;
%mend lifehaz;
```

THE PREDICT MACRO

The PREDICT macro produces predicted survival probabilities for all individuals in a sample for specified survival times, based on models fitted by PROC LIFEREG. You must specify OUTEST=*name1* in the PROC LIFEREG statement. You must also use the OUTPUT statement with OUT=*name2* and XBETA=*name3*. The probabilities are written to the output window and also to a temporary SAS data set named _PRED_.

To use the macro, you must first read it into your SAS session. This can be done by simply copying the text of the macro into the editor window and then running it. Alternatively, you can use the %INCLUDE statement to specify the name and location of the file containing the macro (which must have a .sas extension). Once the macro has been read in to SAS, you can use it repeatedly during your SAS session. For an example using the PREDICT macro, see the section **Generating Predictions and Hazard Functions** in Chapter 4.

Parameters

OUTEST=	specifies the name of the data set produced by the OUTEST= option in PROC LIFEREG.
OUT=_LAST_	specifies the name of the data set produced by the OUT= option in PROC LIFEREG.
XBETA=	specifies the name used with the XBETA= option in the OUTPUT statement.
TIME=	specifies the survival time to be evaluated. This time must be in the same metric as the event times.

Program

```
%macro predict (outest=, out=_last_,xbeta=,time=);
/*******************************************************************

Example:  To get 5-year survival probabilities for every individual
in the sample (assuming that actual survival times are measured in
years);

%predict(outest=a, out=b, xbeta=lp, time=5).

********************************************************************/
```

```
data _pred_;
_p_=1;
set &outest  point=_p_;
set &out;
lp=&xbeta;
t=&time;
gamma=1/_scale_;
alpha=exp(-lp*gamma);
prob=0;
_dist_=upcase(_dist_);
if _dist_='WEIBULL' or _dist_='EXPONENTIAL' or _dist_='EXPONENT' then prob=exp(-
alpha*t**gamma);
if _dist_='LOGNORMAL' or _dist_='LNORMAL' then prob=1-probnorm((log(t)-lp)/_scale_);
if _dist_='LLOGISTIC' or _dist_='LLOGISTC' then prob=1/(1+alpha*t**gamma);
if _dist_='GAMMA' then do;
  d=_shape1_;
  k=1/(d*d);
  u=(t*exp(-lp))**gamma;
  prob=1-probgam(k*u**d,k);
  if d lt 0 then prob=1-prob;
  end;
drop lp gamma alpha _dist_ _scale_ intercept
     _shape1_ _model_ _name_ _type_ _status_ _prob_ _lnlike_ d k u;
run;
proc print data=_pred_;
run;
%mend predict;
```

298

APPENDIX **2**
Data Sets

INTRODUCTION

This appendix provides information about the data sets used as examples in this book. They are listed in the order in which they are discussed in the main text. All data are written in free format, with the variables appearing in the order given in each of the following sections. These data sets are available on the Web at support.sas.com/publishing/authors/allison.html.

THE MYEL DATA SET: MYELOMATOSIS PATIENTS

The MYEL data set contains survival times for 25 patients diagnosed with myelomatosis (Peto et al., 1977). The patients were randomly assigned to two drug treatments. The entire data set is listed in Output 2.1. These data are used extensively in Chapter 3 to illustrate the LIFETEST procedure and briefly in Chapter 5. The variables are as follows:

DUR   is the time in days from the point of randomization to either death or censoring (which could occur either by loss to follow up or termination of the observation).

STATUS   has a value of 1 if dead and a value of 0 if censored.

TREAT   specifies a value of 1 or 2 to correspond to the two treatments.

RENAL   has a value of 1 if renal functioning was normal at the time of randomization; it has a value of 0 for impaired functioning.

THE RECID DATA SET: ARREST TIMES FOR RELEASED PRISONERS

The RECID data set contains information about 432 inmates who were released from Maryland state prisons in the early 1970s. This data set is used in Chapters 3, 4, and 5. The first 20 cases are listed in Output 3.10. The data were kindly provided to me by Dr. Kenneth Lenihan who was one of the principal investigators. The aim of this research was to determine the efficacy of financial aid to released inmates as a means of reducing recidivism. Results from the study are described in Rossi, Berk, and Lenihan (1980). (This book also reports results from a much larger follow-up study done in Texas and Georgia.) Half the inmates were randomly assigned to receive financial aid (approximately the same amount as unemployment compensation). They were followed for 1 year after their release and were interviewed monthly during that period. Data on arrests were taken from police and court records. The data set used here contains the following variables:

WEEK   is the week of first arrest; WEEK has a value of 52 if not arrested.

ARREST   has a value of 1 if arrested; otherwise, ARREST has a value of 0.

FIN   has a value of 1 if the inmate received financial aid after release; otherwise, FIN has a value of 0. FIN is randomly assigned, with equal numbers in each category.

AGE   is the age in years at the time of release.

RACE   has a value of 1 if the inmate is black; otherwise, RACE has a value of 0.

WEXP   has a value of 1 if the inmate had full-time work experience before incarceration; otherwise, WEXP has a value of 0.

MAR   has a value of 1 if the inmate was married at the time of release; otherwise, MAR has a value of 0.

PARO   has a value of 1 if released on parole; otherwise, PARO has a value of 0.

PRIO   is the number of convictions before the current incarceration.

EDUC is the highest level of completed schooling, coded as
 2 = 6th grade or less
 3 = 7th to 9th grade
 4 = 10th to 11th grade
 5 = 12th grade
 6 = some college

EMP1–EMP52 represents the employment status in each of the first 52 weeks after release. These variables have values of 1 if the inmate was employed full time; otherwise, EMP1–EMP52 have values of 0. Data are missing for weeks after the first arrest.

THE STAN DATA SET: STANFORD HEART TRANSPLANT PATIENTS

Various versions of data from the Stanford Heart Transplant study have been reported in a number of publications. The data set used in Chapter 5, "Estimating Cox Regression Models with PROC PHREG," appeared in Crowley and Hu (1977). The sample consisted of 103 cardiac patients who were enrolled in the transplantation program between 1967 and 1974. After enrollment, patients waited varying lengths of time until a suitable donor heart was found. Patients were followed until death or until the termination date of April 1, 1974. Of the 69 transplant recipients, only 24 were still alive at termination. At the time of transplantation, all but four of the patients were tissue typed to determine the degree of similarity with the donor. The data set contains the following variables:

DOB is the date of birth.
DOA is the date of acceptance into the program.
DOT is the date of transplant.
DLS is the date last seen (death date or censoring date).
DEAD has a value of 1 if the patient is dead at DLS; otherwise, DEAD has a value of 0.
SURG has a value of 1 if the patient had open-heart surgery before DOA; otherwise, SURG has a value of 0.
M1 is the number of donor alleles with no match in the recipient (1 through 4).
M2 has a value of 1 if the donor and recipient mismatch on the HLA-A2 antigen; otherwise, M2 has a value of 0.
M3 is the mismatch score.

All date variables are in the *mm/dd/yy* format.

THE BREAST DATA SET: SURVIVAL DATA FOR BREAST CANCER PATIENTS

The BREAST data set (displayed in Output 5.4) contains survival data for 45 breast cancer patients. The data are taken from Collett (1994) except that survival time for the 8th patient is changed from 26 to 25 to eliminate ties. The variables are as follows:

SURV   is the survival time (or censoring time) in months, beginning with the month of surgery.

DEAD   1 = dead; 0=alive.

X   has a value of 1 if the tumor had a positive marker for metastasis; otherwise, X has a value of 0.

THE JOBDUR DATA SET: DURATIONS OF JOBS

The JOBDUR data set consists of 100 simulated observations of jobs and job holders. Results of analyzing these data were reported in Output 5.9 and 5.10. The variables are as follows:

DUR   is the length of the job in years (integer values only).

EVENT   1 = quit; 2 = fired; 0 = censored.

ED   is the number of years of completed schooling.

PRESTIGE   is a measure of the prestige of the job.

SALARY   is the salary in thousands of dollars.

For the analyses in Chapter 5, firings were treated as censoring.

THE ALCO DATA SET: SURVIVAL OF CIRRHOSIS PATIENTS

The ALCO data set, displayed in Output 5.15, consists of simulated data for 29 alcoholic cirrhosis patients. The variables are as follows:

SURV   is the number of months from diagnosis until death or termination of the study.

DEAD   has a value of 1 if the patient died; otherwise, DEAD has a value of 0.

TIME2–TIME10   represent the time of clinic visits in months since diagnosis.

PT1–PT10   is the blood coagulation measure at diagnosis and at subsequent clinic visits.

THE LEADERS DATA SET: TIME IN POWER FOR LEADERS OF COUNTRIES

The LEADERS data set, analyzed in Chapter 6, "Competing Risks," is described in detail by Henry Bienen and Nicolas van de Walle (1991), who generously provided me with the data. Each record in the data set corresponds to a spell in power for a primary leader of a country. I used a subset of the original data, restricting the spells to countries outside of Europe, North America, and Australia; spells that began in 1960 or later; and only the first leadership spell for those leaders with multiple spells. This left a total of 472 spells.

Each record includes the following variables:

YEARS   is the number of years in power, integer valued. Leaders in power less than 1 year have a value of 0.

LOST   0 = still in power in 1987; 1 = exit by constitutional means; 2 = death by natural causes; and 3 = nonconstitutional exit.

MANNER   specifies how the leader reached power: 0 = constitutional means; 1 = nonconstitutional means.

START   is the leader's year of entry into power.

MILITARY   is the background of the leader: 1 = military; 0 = civilian.

AGE   is the age of the leader, in years, at the time of entry into power.

CONFLICT   is the level of ethnic conflict: 1 = medium or high; 0 = low.

LOGINC   is the natural logarithm of GNP per capita (dollar equivalent) in 1973.

GROWTH   is the average annual rate of per capita GNP growth between 1965 and 1983.

POP   is the population, in millions (year not indicated).

LAND   is the land area, in thousands of square kilometers.

LITERACY   is the literacy rate (year not indicated).

REGION   0 = Middle East; 1 = Africa; 2 = Asia; 3 = Latin America.

THE RANK DATA SET: PROMOTIONS IN RANK FOR BIOCHEMISTS

The RANK data set, analyzed in Chapter 7, "Analysis of Tied or Discrete Data with PROC LOGISTIC," contains records for 301 biochemists who at some point in their careers were assistant professors at graduate departments in the U.S. The event of interest is a promotion to associate professor. The data set includes the following variables:

DUR
: is the number of years from the beginning of the job to promotion or censoring (ranges from 1 to 10, in integers).

EVENT
: has a value of 1 if the person was promoted; otherwise, EVENT has a value of 0.

UNDGRAD
: specifies the selectivity of the undergraduate institution (ranges from 1 to 7).

PHDMED
: has a value of 1 if the person received a Ph.D. from a medical school; otherwise, PHDMED has a value of 0.

PHDPREST
: is a measure of the prestige of the Ph.D. institution (ranges from 0.92 to 4.62).

ART1–ART10
: specify the cumulative number of articles published in each of the 10 years.

CIT1–CIT10
: specify the number of citations in each of the 10 years to all previous articles.

PREST1
: is a measure of the prestige of the first employing institution (ranges from 0.65 to 4.6).

PREST2
: is the measure of the prestige of the second employing institution (same as PREST1 for those who did not change employers). No one had more than two employers during the period of observation.

JOBTIME
: is the year of employer change, measured from the start of the assistant professorship (coded as missing for those who did not change employers).

THE JOBMULT DATA SET: REPEATED JOB CHANGES

The JOBMULT data set contains simulated information for 395 jobs observed over a period of 20 years for 100 people. The first 20 cases are displayed in Output 8.1, with analytic results in Output 8.2 through 8.9.

ID   is a unique identification number for each *person*.

EVENT   has a value of 1 if the job was terminated; EVENT has a value of 0 if censored.

ED   is the number of years of schooling completed.

SEQ   is the number of the job in each person's sequence.

DURATION   is the length of the job until termination or censoring.

PRESTIGE   is a measure of the prestige of the job, ranging from 1 to 100.

LOGSAL   is the natural logarithm of the annual salary at the beginning of the job.

306

REFERENCES

Aitkin, M.; Anderson, D. A.; Francis, B.; and Hinde, J. P. (1989), *Statistical Modelling in GLIM,* Oxford: Clarendon Press.

Allison, P. D. (1982), "Discrete-Time Methods for the Analysis of Event Histories," in *Sociological Methodology 1982*, ed. S. Leinhardt, San Francisco, CA: Jossey-Bass, 61–98.

Allison, P. D. (1984), *Event History Analysis*, Beverly Hills, CA: Sage Publications.

Allison, P. D. (1987), "Introducing a Disturbance Into Logit and Probit Regression Models," *Sociological Methods and Research,* 15, 355–374.

Allison, P. D. (1996), "Fixed Effects Partial Likelihood for Repeated Events," *Sociological Methods and Research*, 25, 207–222.

Allison, P. D. (2002), "Bias in Fixed-Effects Cox Regression with Dummy Variables," unpublished paper, Department of Sociology, University of Pennsylvania.

Andersen, P. K. and Gill, R. D. (1982), "Cox's Regression Model for Counting Processes: A Large Sample Study," *The Annals of Statistics*, 10, 1100–1120.

Begg, C. B. and Gray, R. (1984), "Calculation of Polychotomous Logistic Regression Parameters Using Individualized Regressions," *Biometrika*, 71, 11–18.

Bienen, H. S. and van de Walle, N. (1991), *Of Time and Power*, Stanford, CA: Stanford University Press.

Breslow, N. E. (1974), "Covariance Analysis of Censored Survival Data," *Biometrics*, 30, 89–99.

308

Chamberlain, G. A. (1985), "Heterogeneity, Omitted Variable Bias, and Duration Dependence," in *Longitudinal Analysis of Labor Market Data*, eds. J. J. Heckman and B. Singer, New York: Cambridge University Press, 3–38.

Collett, D. (2003), *Modelling Survival Data in Medical Research*, Second Edition, London: Chapman & Hall.

Cox, D. R. (1972), "Regression Models and Life Tables" (with discussion), *Journal of the Royal Statistical Society,* B34, 187–220.

Cox, D. R. and Oakes, D. (1984), *Analysis of Survival Data*, London: Chapman & Hall.

Cox, D. R. and Snell, E. J. (1989), *The Analysis of Binary Data*, Second Edition, London: Chapman & Hall.

Crowley, J. and Hu, M. (1977), "Covariance Analysis of Heart Transplant Survival Data," *Journal of the American Statistical Association*, 72, 27–36.

Daniels, M. J. and Hogan, J. W. (2008), *Missing Data in Longitudinal Studies: Strategies for Bayesian Modeling and Sensitivity Analysis*, London: Chapman and Hall.

DeLong, D. M.; Guirguis, G. H.; and So, Y. C. (1994), "Efficient Computation of Subset Selection Probabilities with Application to Cox Regression," *Biometrika*, 81, 607–611.

Efron, B. (1977), "The Efficiency of Cox's Likelihood Function for Censored Data," *Journal of the American Statistical Association*, 76, 312–319.

Elandt-Johnson, R. C. and Johnson, N. L. (1980), *Survival Models and Data Analysis*, New York: John Wiley & Sons, Inc.

Farewell, V. T. and Prentice, R. L. (1977), "A Study of Distributional Shape in Life Testing," *Technometrics*, 19, 69–75.

Farewell, V. T. and Prentice, R. L. (1980), "The Approximation of Partial Likelihood with Emphasis on Case-Control Studies," *Biometrika*, 67, 273–278.

Fine, J. P., Yan, J. and Kosorok, M. R. (2004), "Temporal Process Regression," *Biometrika* 91, 683–703.

Firth, D. (1993), "Bias Reduction of Maximum Likelihood Estimates," *Biometrika*, 80, 27–38.

Gail, M. H.; Lubin, J. H.; and Rubinstein, L. V. (1981), "Likelihood Calculations for Matched Case-Control Studies and Survival Studies with Tied Death Times," *Biometrika*, 68, 703–707.

Gail, M. H.; Wieand, S.; and Piantadosi, S. (1984), "Biased Estimates of Treatment Effect in Randomized Experiments with Nonlinear Regressions and Omitted Covariates," *Biometrika*, 71, 431–444.

Garfield, E. (1990), "100 Most Cited Papers of All Time," *Current Contents*, February 12.

Gray, R. J. (1988), "A Class of K-Sample Tests for Comparing the Cumulative Incidence of a Competing Risk," *Annals of Statistics*, 16, 1141–54.

Gross, A. J. and Clark, V. A. (1975), *Survival Distributions: Reliability Applications in the Biomedical Sciences*, New York: John Wiley & Sons, Inc.

Hannan, M. T. and Freeman, J. (1984), "Structural Inertia and Organizational Change," *American Sociological Review*, 49, 149–164.

Harris, E. K. and Albert, A. (1991), *Survivorship Analysis for Clinical Studies*, New York: Marcel Dekker, Inc.

Heckman, J. J. and Honoré, B. E. (1989), "The Identifiability of the Competing Risks Model," *Biometrika*, 76, 325–330.

Heckman, J. J. and Singer, B. (1985), "Social Science Duration Analysis" in *Longitudinal Studies of Labor Market Data*, ed. J. J. Heckman and B. Singer, New York: Cambridge University Press, Chapter 2.

Heinze, Georg and Schemper, Michael (2001), "A Solution to the Problem of Monotone Likelihood in Cox Regression," *Biometrics*, 57, 114–119.

Hsieh, Frank Y. (1995), "A Cautionary Note on the Analysis of Extreme Data with Cox Regression," *The American Statistician*, 49, 226–228.

Institute for Scientific Information (1992), *Science Citation Index*, Philadelphia: Institute for Scientific Information.

Kalbfleisch, J. D. and Prentice, R. L. (2002), *The Statistical Analysis of Failure Time Data*, Second Edition, New York: John Wiley & Sons, Inc.

Kaplan, E. L. and Meier, P. (1958), "Nonparametric Estimation from Incomplete Observations," *Journal of the American Statistical Association*, 53, 457–481.

Klein, J. P. (1992), "Semiparametric Estimation of Random Effects Using the Cox Model Based on the EM Algorithm," *Biometrics*, 48, 795–806.

Lagakos, S. W. (1978), "A Covariate Model for Partially Censored Data Subject to Competing Causes of Failure," *Applied Statistics*, 27, 235–241.

Lawless, J. F. (2002), *Statistical Models and Methods for Lifetime Data*, Second Edition, New York: John Wiley & Sons, Inc.

Lee, E. T. (1992), *Statistical Methods for Survival Data Analysis*, Second Edition, New York: John Wiley & Sons, Inc.

Long, J. S.; Allison, P. D.; and McGinnis, R. (1993), "Rank Advancement in Academic Careers: Sex Differences and the Effects of Productivity," *American Sociological Review*, 58, 703–722.

Magee, L. (1990), "R^2 Measures Based on Wald and Likelihood Ratio Joint Significance Tests," *The American Statistician*, 44, 250–253.

Marubini, E. and Valsecchi, M. G. (1995), *Analysing Survival Data from Clinical Trials and Observational Studies*, New York: John Wiley & Sons, Inc.

McGilchrist, C. A. (1993), "REML Estimation for Survival Models with Frailty," *Biometrics*, 49, 221–225.

Nakamura, T. (1992), "Proportional Hazards Model with Covariates Subject to Measurement Error," *Biometrics*, 48, 829–838.

Narendranathan, W. and Stewart, M. B. (1991), "Simple Methods for Testing for the Proportionality of Cause-Specific Hazards in Competing Risks Models," *Oxford Bulletin of Economics and Statistics*, 53, 331–340.

Petersen, T. (1991), "Time Aggregation Bias in Continuous-Time Hazard Rate Models," in *Sociological Methodology 1991*, ed. P. V. Marsden, Oxford: Basil Blackwell, 263–290.

Pintilie, M. (2006), *Competing Risks: A Practical Perspective,* New York: John Wiley & Sons, Inc.

Peto, R. (1972), Contribution to the discussion of paper by D. R. Cox," *Journal of the Royal Statistical Society*, B34, 205–207.

Peto, R.; Pike, M. C.; Armitage, P.; Breslow, N. E.; Cox, D. R.; Howard, S. V.; Mantel, N.; McPherson, K.; Peto, J.; and Smith, P. G. (1977), "Design and Analysis of Randomized Clinical Trials Requiring Prolonged Observation of Each Patient. II. Analysis and Examples," *British Journal of Cancer*, 35, 1–39.

Prentice, R. L. and Gloeckler, L. A. (1978), "Regression Analysis of Grouped Survival Data with Application to Breast Cancer Data," *Biometrics*, 34, 57–67.

Prentice, R. L., Williams, B. J., and Peterson, A. V. (1981), "On the Regression Analysis of Multivariate Failure Time Data," *Biometrika*, 68, 373–379.

Prentice, R. L. and Pike, R. (1979), "Logistic Disease Incidence Models and Case-Control Studies," *Biometrika*, 66, 403–411.

Proschan, F. (1963), "Theoretical Explanation of Observed Duration Dependence," *Technometrics*, 5, 375–383.

Ramlau-Hansen, H. (1983), "Smoothing Counting Process Intensities by Means of Kernel Functions," *The Annals of Statistics,* 11, 453–466.

Rossi, P. H.; Berk, R. A.; and Lenihan, K. J. (1980), *Money, Work and Crime: Some Experimental Results*, New York: Academic Press, Inc.

Scheike, T. H.; Zhang, M.; and Gerds, T.A. (2008), "Predicting Cumulative Incidence Probability by Direct Binomial Regression," *Biometrika,* 95, 205–220.

Thompson, W. A. (1977), "On the Treatment of Grouped Observations in Life Studies," *Biometrics*, 33, 463–470.

Tuma, N. B. (1984), *Invoking RATE.* Unpublished program manual. Stanford, CA: Stanford University Press.

Wei, L. J.; Lin, D. Y.; and Weissfeld, L. (1989), "Regression Analysis of Multivariate Incomplete Failure Time Data by Modeling Marginal Distributions," *Journal of the American Statistical Association*, 84, 1065–1073.

Index

www.ingramcontent.com/pod-product-compliance
Lightning Source LLC
Chambersburg PA
CBHW061323190326
41458CB00011B/3874